Domain	Center	Skill	Page
Numbers and Counting	**Count to 1,000**	Count, write, and sequence numbers within 1,000	5
	Hundreds, Tens, and Ones	Know place values in three-digit numbers	19
	Greater, Less, or Equal?	Compare numbers within 1,000, using greater than, less than, and equal to symbols	33
	Skip Counting to 1,000	Count by 5s, 10s, and 100s	45
Operations	**Facts to 20**	Build fluency with addition and subtraction facts to 20	57
	Fact Families	Know related addition and subtraction facts within 20	71
	Start at the Ones Place	Use place value to add and subtract	89
Measurement and Data	**Use a Ruler**	Measure length in standard units (inches and centimeters)	101
	How Much Money?	Solve word problems, using coins and bills from one cent to one dollar	113
	What Time Is It?	Tell time to the nearest 5 minutes (analog and digital)	123
	Graphs and Data	Read and use data on bar graphs and picture graphs	137
Geometry	**Shapes and Fractions**	Identify fractional parts ($\frac{1}{2}$, $\frac{1}{3}$, $\frac{1}{4}$) of shapes with 3 to 6 sides	149

Using the Centers

The 12 centers in this book provide hands-on practice to help students master standards-based mathematics skills. It is important to teach each skill and to model the use of each center before asking students to do the tasks independently. The centers are self-contained and portable. Students can work at a desk, at a table, or on a rug, and they can use the centers as often as needed.

Why Use Centers?

- Centers are a motivating way for students to practice important skills.
- They provide for differentiated instruction.
- They appeal especially to kinesthetic and visual learners.
- They are ready to use whenever instruction or practice in the target skill is indicated.

Before Using Centers

You and your students will enjoy using centers more if you think through logistical considerations. Here are a few questions to resolve ahead of time:

- Will students select a center, or will you assign the centers and use them as a skill assessment tool?
- Will there be a specific block of time for centers, or will the centers be used by students throughout the day as they complete other work?
- Where will you place the centers for easy access by students?
- What procedure will students use when they need help with the center tasks?
- Will students use the answer key to check their own work?
- How will you use the center checklist to track completion of the centers?

Introducing the Centers

Use the teacher instructions page and the student directions on the center's cover page to teach or review the skill. Show students the pieces of the center and model how to use them as you read each step of the directions.

Some centers have built-in scaffolding, offering more than one level of practice for the target skill. Carefully review each center beforehand to be sure that you are introducing or assigning the most appropriate level for each student or group.

Recording Progress

Use the center checklist (page 4) to record both the date when a student completes each center and the student's skill level at that point.

Making the Centers

Included for Each Center

- (A) Student directions/cover page
- (B) Task cards and mat(s)
- (C) Reproducible activity
- (D) Answer key

Materials Needed

- Folders with inside pockets
- Small envelopes or self-closing plastic bags (for storing task cards)
- Pencils or marking pens (for labeling envelopes)
- Scissors
- Double-sided tape (for attaching the cover page to the front of the folder)
- Laminating equipment

How to Assemble and Store

1. Tape the center's cover page to the front of the folder.
2. Place reproduced activity pages in the left-hand pocket of the folder.
3. Laminate mats and task cards.
4. Cut apart the task cards and put them in a labeled envelope or self-closing plastic bag. Place the mats and task cards in the right-hand pocket of the folder. If you want the centers to be self-checking, include the answer key in the folder.
5. Store prepared centers in a file box or a crate.

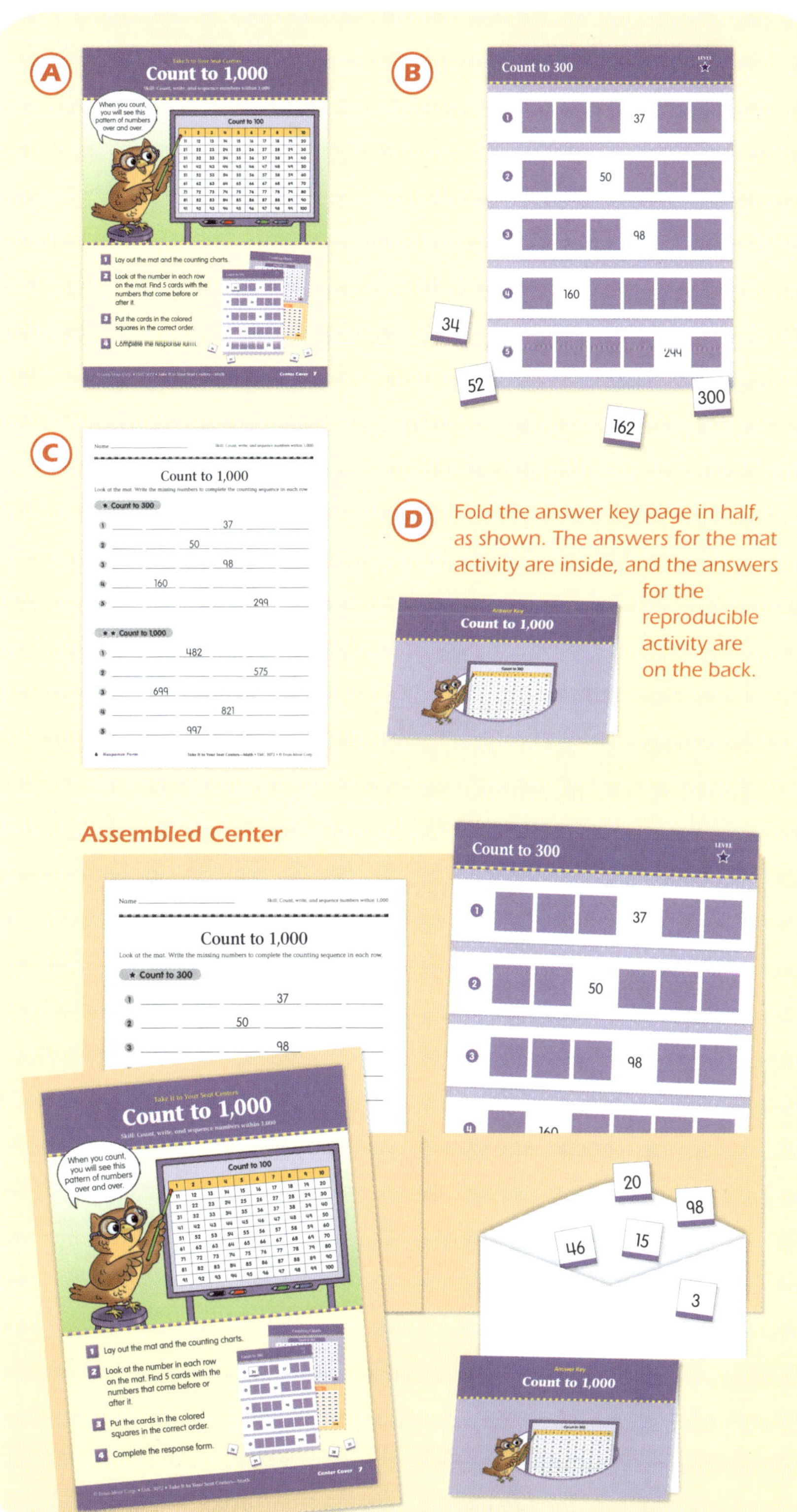

Student ______________________________

Center Checklist

Center / Skill	Skill Level	Date
1. Count to 1,000 Count, write, and sequence numbers within 1,000		
2. Hundreds, Tens, and Ones Know place values in three-digit numbers		
3. Greater, Less, or Equal? Compare numbers within 1,000, using greater than, less than, and equal to symbols		
4. Skip Counting to 1,000 Count by 5s, 10s, and 100s		
5. Facts to 20 Build fluency with addition and subtraction facts to 20		
6. Fact Families Know related addition and subtraction facts within 20		
7. Start at the Ones Place Use place value to add and subtract		
8. Use a Ruler Measure length in standard units (inches and centimeters)		
9. How Much Money? Solve word problems, using coins and bills from one cent to one dollar		
10. What Time Is It? Tell time to the nearest 5 minutes (analog and digital)		
11. Graphs and Data Read and use data on bar graphs and picture graphs		
12. Shapes and Fractions Identify fractional parts ($\frac{1}{2}$, $\frac{1}{3}$, $\frac{1}{4}$) of shapes with 3 to 6 sides		

Take It to Your Seat Centers—Math • EMC 3072 • © Evan-Moor Corp.

Take It to Your Seat Centers

Count to 1,000

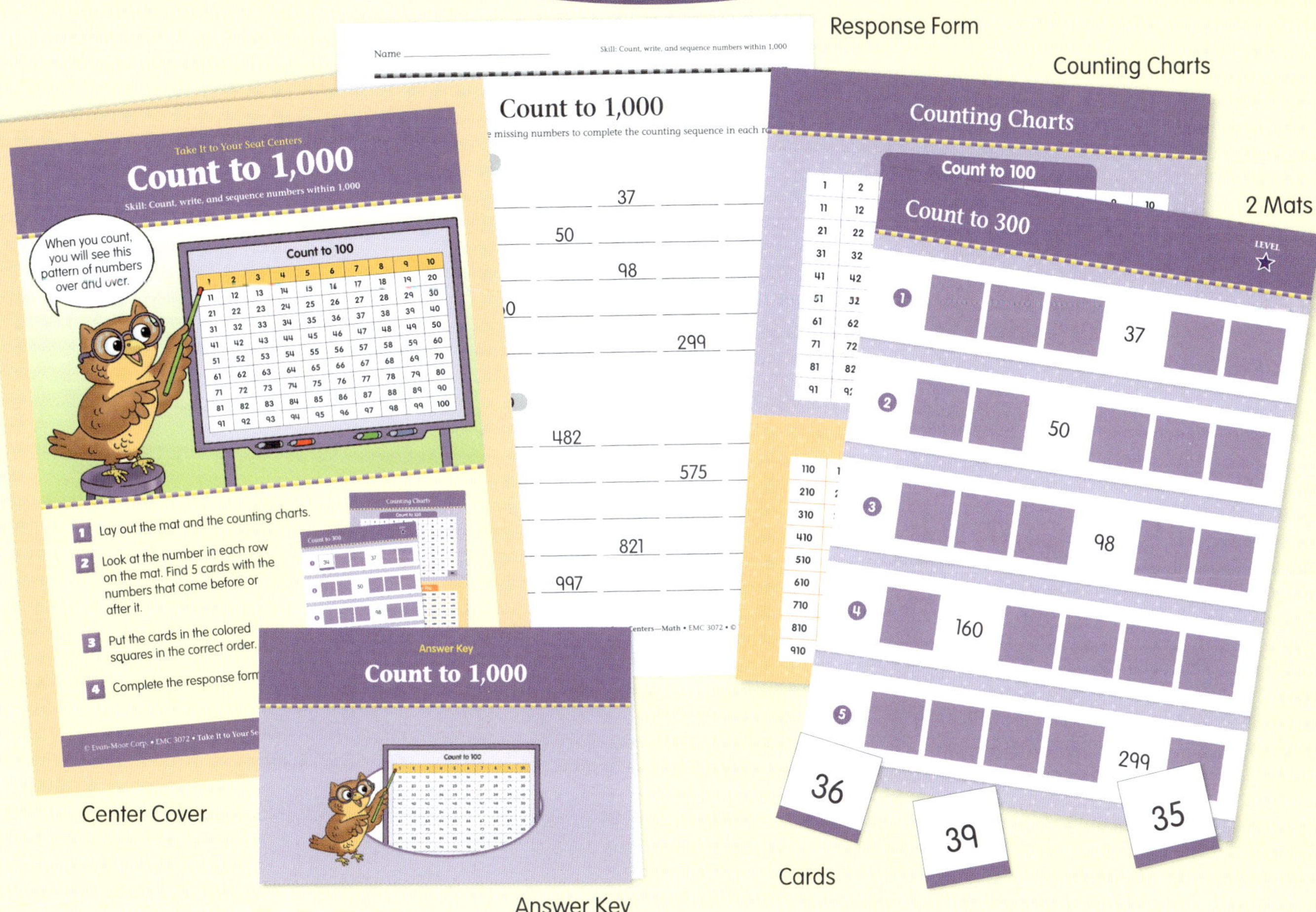

Skill: Count, write, and sequence numbers within 1,000

Steps to Follow

1. **Prepare the center.** (See page 3.)
2. **Introduce the center.** State the goal. Say: *You will place number cards on the mat to complete each counting sequence. You may use the counting charts if you need help.*
3. **Teach the skill.** Demonstrate how to use the center with individual students or small groups.
4. **Practice the skill.** Have students use the center independently or with a partner.

Contents

Response Form 6

Center Cover 7

Answer Key 9

Counting Charts 11

Center Mats

Level 1 13

Level 2 15

Cards 17

Name ______________________ Skill: Count, write, and sequence numbers within 1,000

Count to 1,000

Look at the mat. Write the missing numbers to complete the counting sequence in each row.

★ Count to 300

1 ______ ______ ______ 37 ______ ______

2 ______ ______ 50 ______ ______ ______

3 ______ ______ ______ 98 ______ ______

4 ______ 160 ______ ______ ______ ______

5 ______ ______ ______ ______ 299 ______

★★ Count to 1,000

1 ______ ______ 482 ______ ______ ______

2 ______ ______ ______ ______ 575 ______

3 ______ 699 ______ ______ ______ ______

4 ______ ______ ______ 821 ______ ______

5 ______ ______ 997 ______ ______ ______

Take It to Your Seat Centers

Count to 1,000

Skill: Count, write, and sequence numbers within 1,000

Count to 100

1	2	3	4	5	6	7	8	9	10
11	12	13	14	15	16	17	18	19	20
21	22	23	24	25	26	27	28	29	30
31	32	33	34	35	36	37	38	39	40
41	42	43	44	45	46	47	48	49	50
51	52	53	54	55	56	57	58	59	60
61	62	63	64	65	66	67	68	69	70
71	72	73	74	75	76	77	78	79	80
81	82	83	84	85	86	87	88	89	90
91	92	93	94	95	96	97	98	99	100

1. Lay out the mat and the counting charts.
2. Look at the number in each row on the mat. Find 5 cards with the numbers that come before or after it.
3. Put the cards in the colored squares in the correct order.
4. Complete the response form.

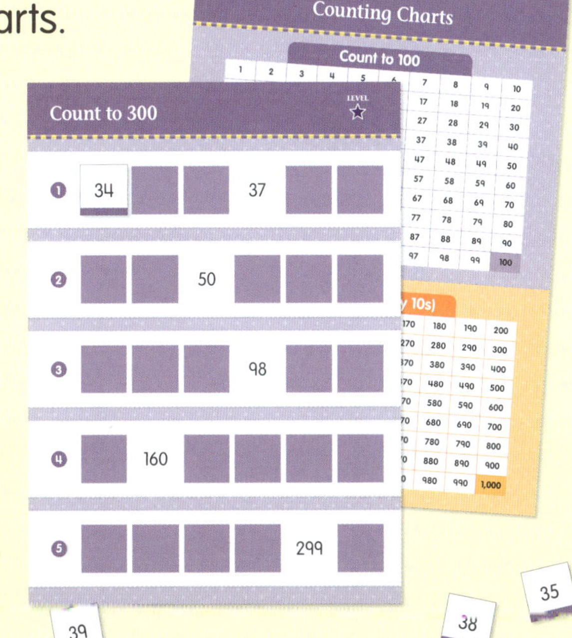

Count to 1,000

Answer Key

(fold)

Response Form

Count to 1,000

Look at the mat. Write the missing numbers to complete the counting sequence in each row.

★ Count to 300

1	34	35	36	37	38	39
2	48	49	50	51	52	53
3	95	96	97	98	99	100
4	159	160	161	162	163	164
5	295	296	297	298	299	300

★★ Count to 1,000

1	480	481	482	483	484	485
2	571	572	573	574	575	576
3	698	699	700	701	702	703
4	818	819	820	821	822	823
5	995	996	997	998	999	1,000

Answer Key

Count to 1,000

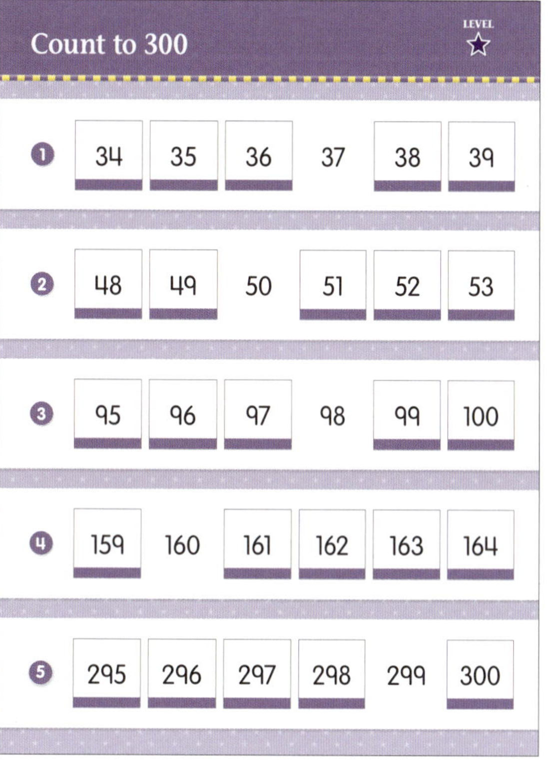

Counting Charts

Count to 100

1	2	3	4	5	6	7	8	9	10
11	12	13	14	15	16	17	18	19	20
21	22	23	24	25	26	27	28	29	30
31	32	33	34	35	36	37	38	39	40
41	42	43	44	45	46	47	48	49	50
51	52	53	54	55	56	57	58	59	60
61	62	63	64	65	66	67	68	69	70
71	72	73	74	75	76	77	78	79	80
81	82	83	84	85	86	87	88	89	90
91	92	93	94	95	96	97	98	99	100

Count to 1,000 (by 10s)

110	120	130	140	150	160	170	180	190	200
210	220	230	240	250	260	270	280	290	300
310	320	330	340	350	360	370	380	390	400
410	420	430	440	450	460	470	480	490	500
510	520	530	540	550	560	570	580	590	600
610	620	630	640	650	660	670	680	690	700
710	720	730	740	750	760	770	780	790	800
810	820	830	840	850	860	870	880	890	900
910	920	930	940	950	960	970	980	990	1,000

Take It to Your Seat Centers—Math • EMC 3072 • © Evan-Moor Corp.

Count to 300

LEVEL

1. ___ ___ ___ 37 ___ ___

2. ___ ___ 50 ___ ___ ___

3. ___ ___ ___ 98 ___ ___

4. ___ 160 ___ ___ ___ ___

5. ___ ___ ___ ___ 299 ___

Count to 1,000

LEVEL ★★

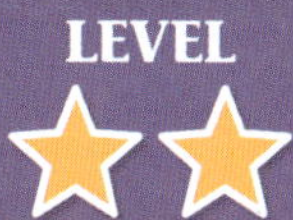

1			482			
2					575	
3		699				
4				821		
5			997			

34	35	36	38	39	48	49
51	52	53	95	96	97	99
100	159	161	162	163	164	295
296	297	298	300			
480	481	483	484	485	571	572
573	574	576	698	700	701	702
703	818	819	820	822	823	995
996	998	999	1,000			

Count to 1,000
EMC 3072
© Evan-Moor Corp.

Count to 1,000
EMC 3072
© Evan-Moor Corp.

Count to 1,000
EMC 3072
© Evan-Moor Corp.

Count to 1,000
EMC 3072
© Evan-Moor Corp.

Count to 1,000
EMC 3072
© Evan-Moor Corp.

Count to 1,000
EMC 3072
© Evan-Moor Corp.

Count to 1,000
EMC 3072
© Evan-Moor Corp.

Count to 1,000
EMC 3072
© Evan-Moor Corp.

Count to 1,000
EMC 3072
© Evan-Moor Corp.

Count to 1,000
EMC 3072
© Evan-Moor Corp.

Count to 1,000
EMC 3072
© Evan-Moor Corp.

Count to 1,000
EMC 3072
© Evan-Moor Corp.

Count to 1,000
EMC 3072
© Evan-Moor Corp.

Count to 1,000
EMC 3072
© Evan-Moor Corp.

Count to 1,000
EMC 3072
© Evan-Moor Corp.

Count to 1,000
EMC 3072
© Evan-Moor Corp.

Count to 1,000
EMC 3072
© Evan-Moor Corp.

Count to 1,000
EMC 3072
© Evan-Moor Corp.

Count to 1,000
EMC 3072
© Evan-Moor Corp.

Count to 1,000
EMC 3072
© Evan-Moor Corp.

Count to 1,000
EMC 3072
© Evan-Moor Corp.

Count to 1,000
EMC 3072
© Evan-Moor Corp.

Count to 1,000
EMC 3072
© Evan-Moor Corp.

Count to 1,000
EMC 3072
© Evan-Moor Corp.

Count to 1,000
EMC 3072
© Evan-Moor Corp.

Count to 1,000
EMC 3072
© Evan-Moor Corp.

Count to 1,000
EMC 3072
© Evan-Moor Corp.

Count to 1,000
EMC 3072
© Evan-Moor Corp.

Count to 1,000
EMC 3072
© Evan-Moor Corp.

Count to 1,000
EMC 3072
© Evan-Moor Corp.

Count to 1,000
EMC 3072
© Evan-Moor Corp.

Count to 1,000
EMC 3072
© Evan-Moor Corp.

Count to 1,000
EMC 3072
© Evan-Moor Corp.

Count to 1,000
EMC 3072
© Evan-Moor Corp.

Count to 1,000
EMC 3072
© Evan-Moor Corp.

Count to 1,000
EMC 3072
© Evan-Moor Corp.

Count to 1,000
EMC 3072
© Evan-Moor Corp.

Count to 1,000
EMC 3072
© Evan-Moor Corp.

Count to 1,000
EMC 3072
© Evan-Moor Corp.

Count to 1,000
EMC 3072
© Evan-Moor Corp.

Count to 1,000
EMC 3072
© Evan-Moor Corp.

Count to 1,000
EMC 3072
© Evan-Moor Corp.

Count to 1,000
EMC 3072
© Evan-Moor Corp.

Count to 1,000
EMC 3072
© Evan-Moor Corp.

Count to 1,000
EMC 3072
© Evan-Moor Corp.

Count to 1,000
EMC 3072
© Evan-Moor Corp.

Count to 1,000
EMC 3072
© Evan-Moor Corp.

Count to 1,000
EMC 3072
© Evan-Moor Corp.

Count to 1,000
EMC 3072
© Evan-Moor Corp.

Count to 1,000
EMC 3072
© Evan-Moor Corp.

Take It to Your Seat Centers

Hundreds, Tens, and Ones

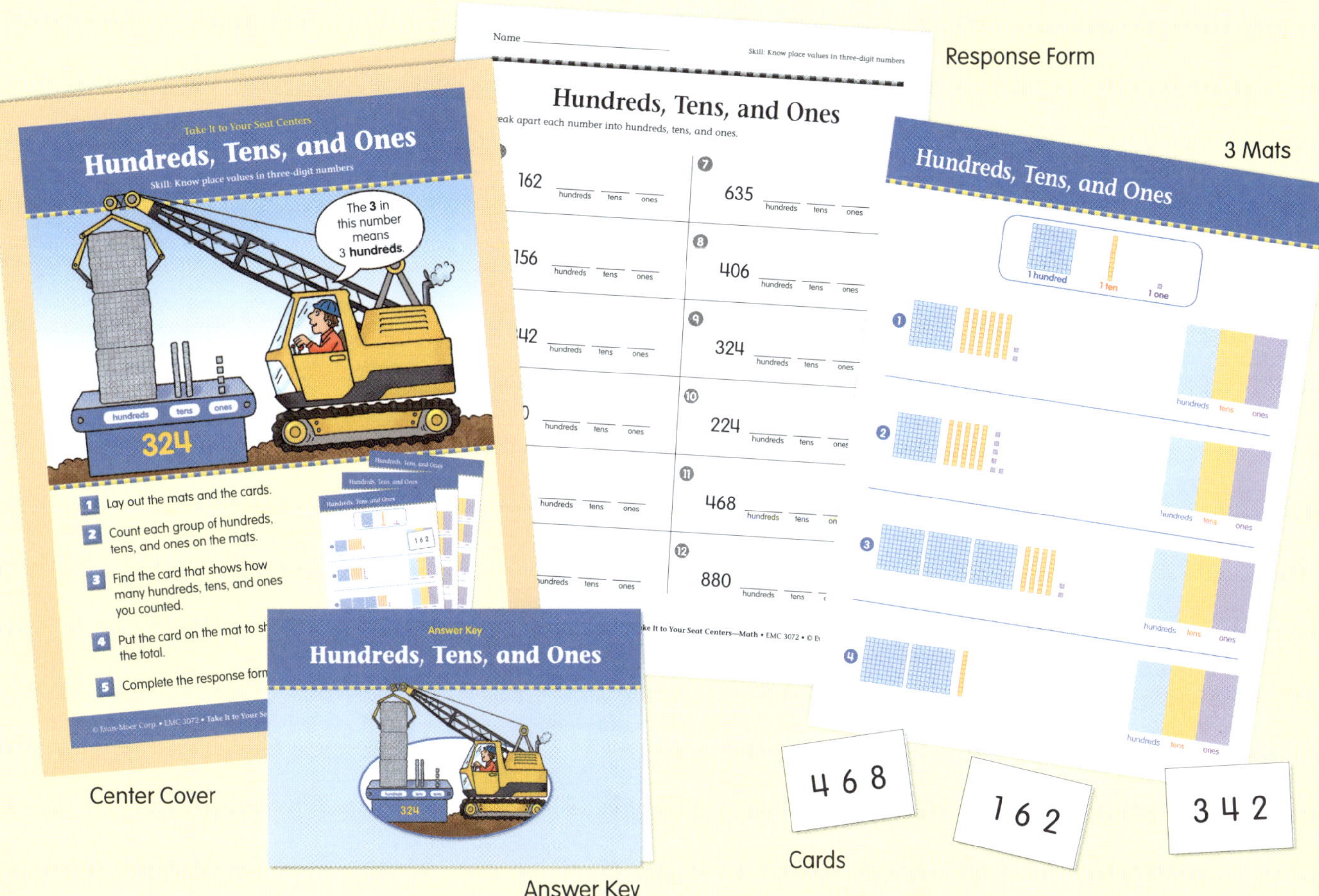

Skill: Know place values in three-digit numbers

Steps to Follow

1. **Prepare the center.** (See page 3.)
2. **Introduce the center.** State the goal. Say: *You will find the card that shows the correct number for each set of blocks on the mats.*
3. **Teach the skill.** Demonstrate how to use the center with individual students or small groups.
4. **Practice the skill.** Have students use the center independently or with a partner.

Contents

Response Form 20
Center Cover 21
Answer Key 23
Center Mats 25, 27, 29
Cards 31

Name ______________________ Skill: Know place values in three-digit numbers

Hundreds, Tens, and Ones

Break apart each number into hundreds, tens, and ones.

1. 162 ____ hundreds ____ tens ____ ones

2. 156 ____ hundreds ____ tens ____ ones

3. 342 ____ hundreds ____ tens ____ ones

4. 210 ____ hundreds ____ tens ____ ones

5. 300 ____ hundreds ____ tens ____ ones

6. 129 ____ hundreds ____ tens ____ ones

7. 635 ____ hundreds ____ tens ____ ones

8. 406 ____ hundreds ____ tens ____ ones

9. 324 ____ hundreds ____ tens ____ ones

10. 224 ____ hundreds ____ tens ____ ones

11. 468 ____ hundreds ____ tens ____ ones

12. 880 ____ hundreds ____ tens ____ ones

Take It to Your Seat Centers

Hundreds, Tens, and Ones

Skill: Know place values in three-digit numbers

The **3** in this number means 3 **hundreds**.

hundreds tens ones

324

1. Lay out the mats and the cards.
2. Count each group of hundreds, tens, and ones on the mats.
3. Find the card that shows how many hundreds, tens, and ones you counted.
4. Put the card on the mat to show the total.
5. Complete the response form.

Response Form

Hundreds, Tens, and Ones

Break apart each number into hundreds, tens, and ones.

#	Number	hundreds	tens	ones
1	162	1	6	2
2	156	1	5	6
3	342	3	4	2
4	210	2	1	0
5	300	3	0	0
6	129	1	2	9
7	635	6	3	5
8	406	4	0	6
9	324	3	2	4
10	224	2	2	4
11	468	4	6	8
12	880	8	8	0

(fold)

Answer Key

Hundreds, Tens, and Ones

Answer Key

Hundreds, Tens, and Ones

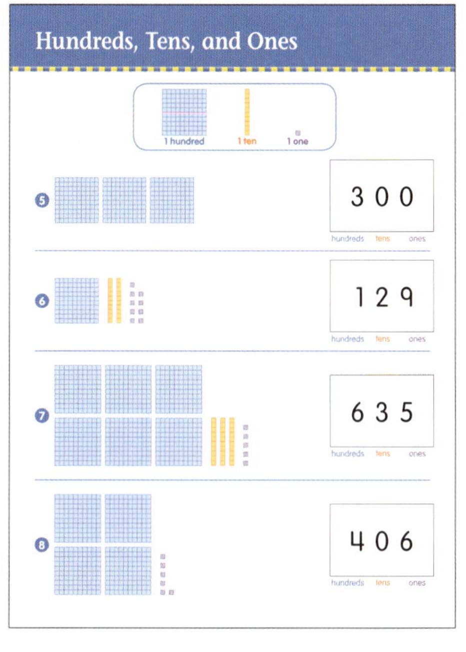

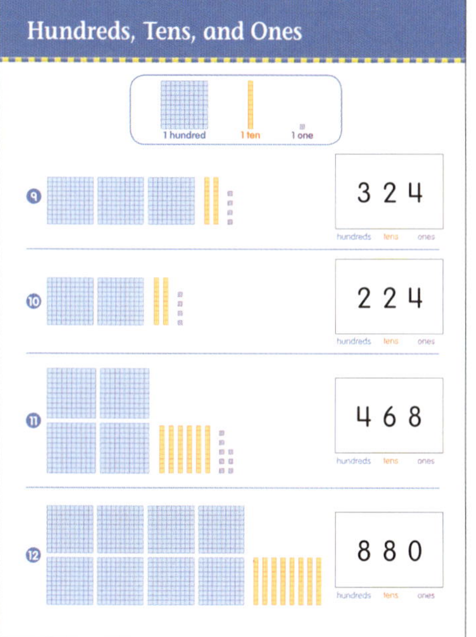

Hundreds, Tens, and Ones

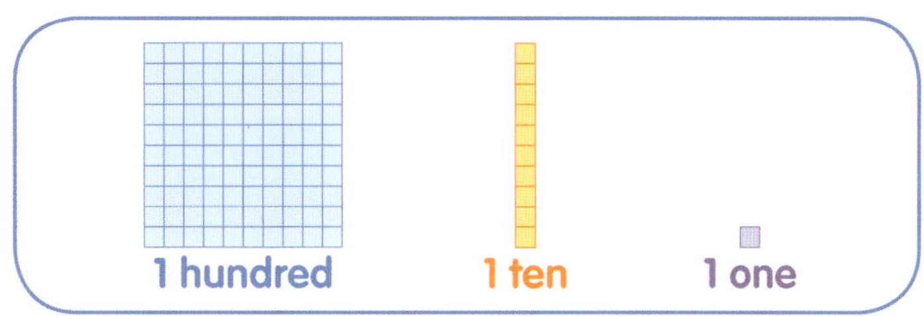

1 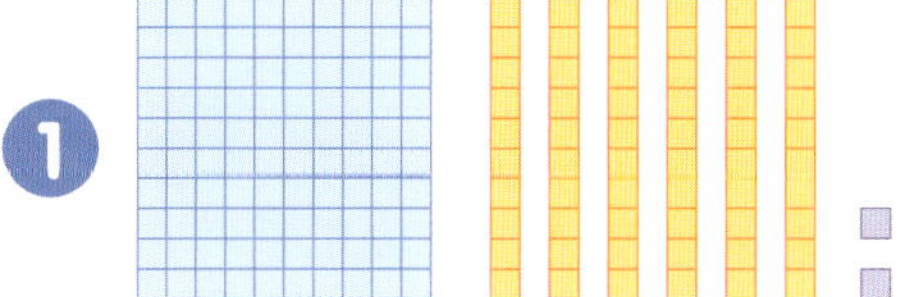

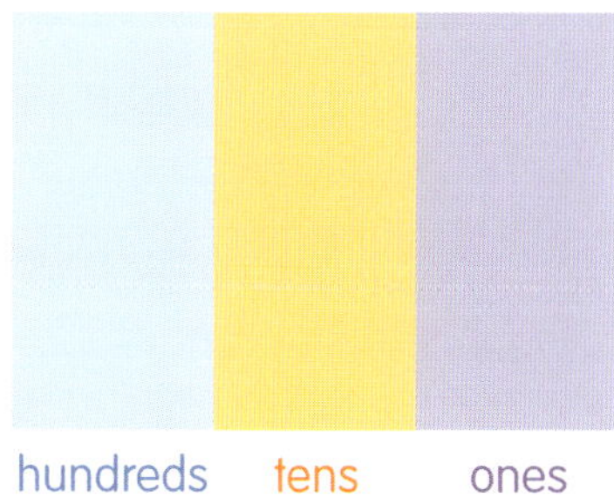

2 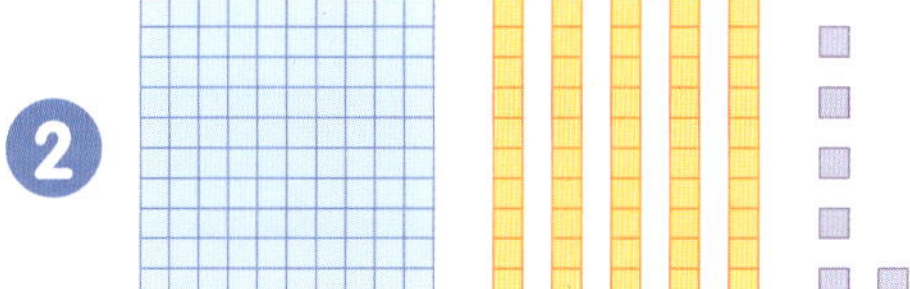

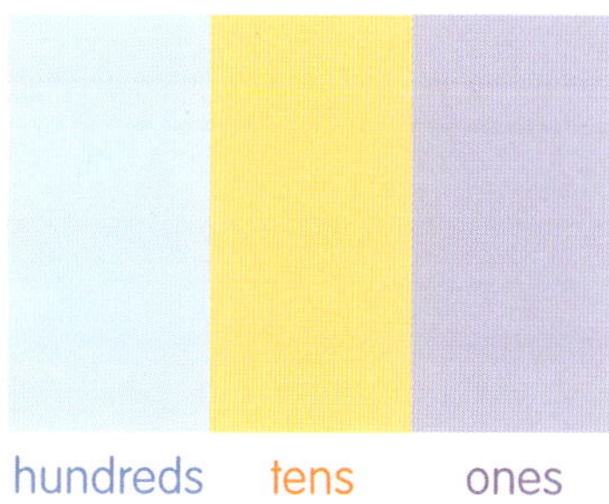

3 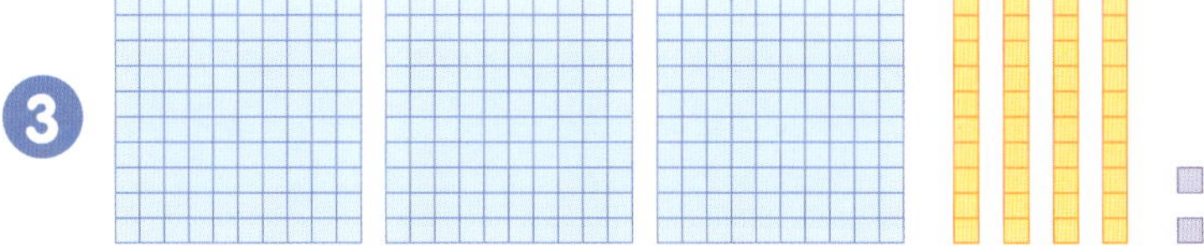

4 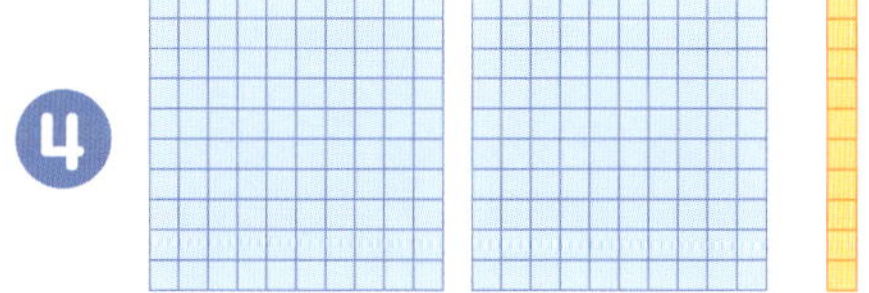

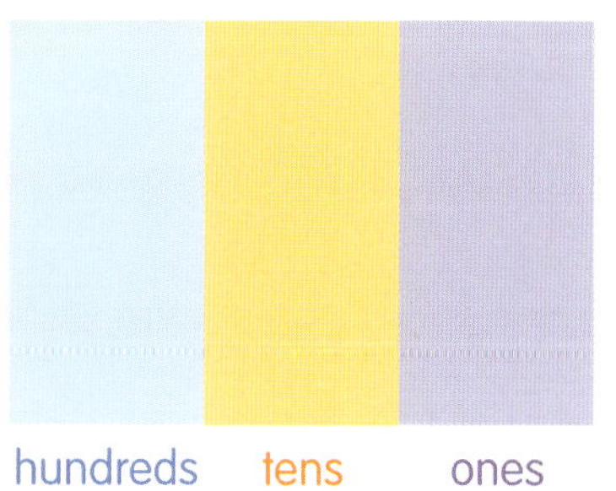

Hundreds, Tens, and Ones

Hundreds, Tens, and Ones

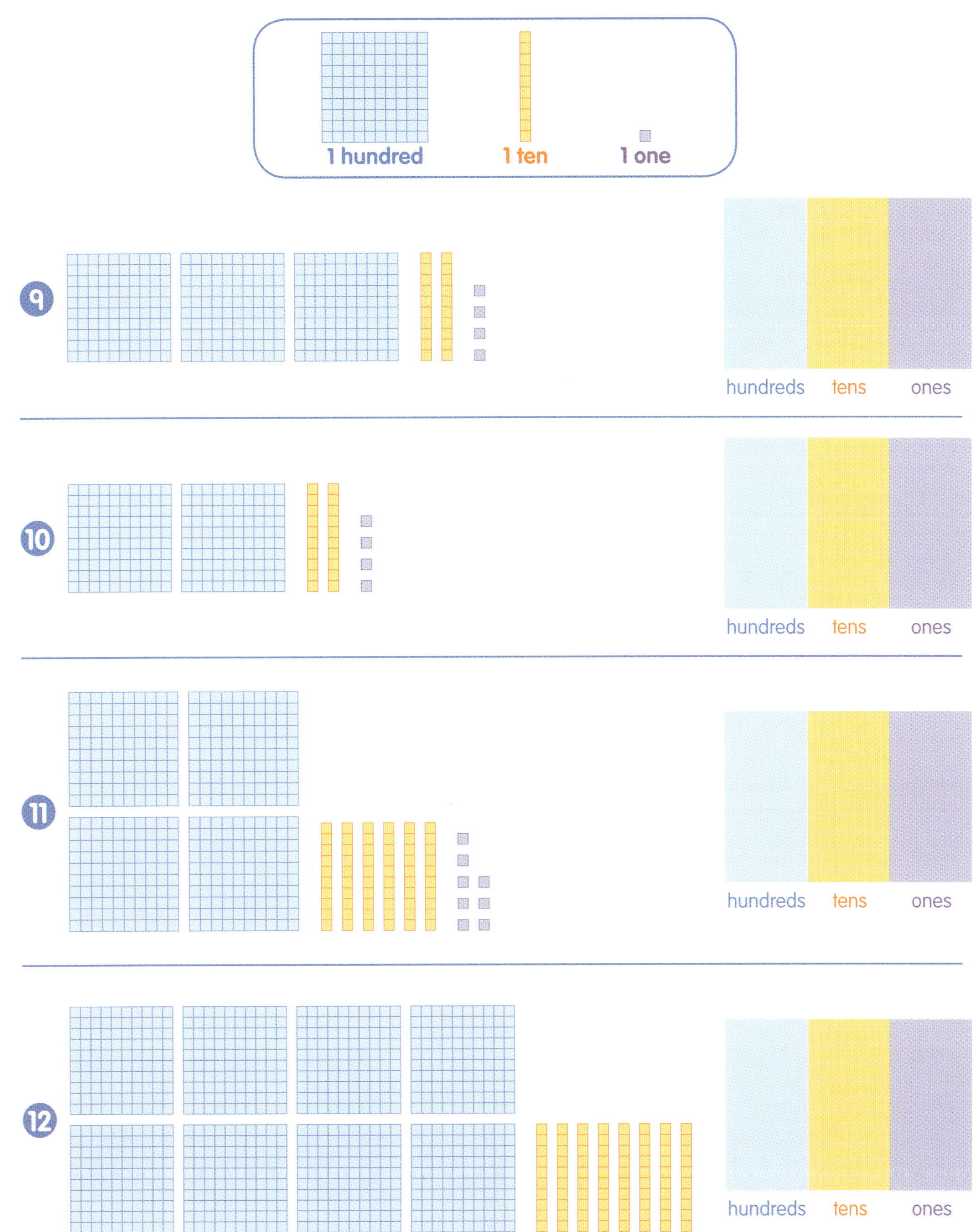

Take It to Your Seat Centers—Math • EMC 3072 • © Evan-Moor Corp.

162	324	468
156	224	342
635	210	880
300	129	406

Hundreds, Tens, and Ones
EMC 3072
© Evan-Moor Corp.

Hundreds, Tens, and Ones
EMC 3072
© Evan-Moor Corp.

Hundreds, Tens, and Ones
EMC 3072
© Evan-Moor Corp.

Hundreds, Tens, and Ones
EMC 3072
© Evan-Moor Corp.

Hundreds, Tens, and Ones
EMC 3072
© Evan-Moor Corp.

Hundreds, Tens, and Ones
EMC 3072
© Evan-Moor Corp.

Hundreds, Tens, and Ones
EMC 3072
© Evan-Moor Corp.

Hundreds, Tens, and Ones
EMC 3072
© Evan-Moor Corp.

Hundreds, Tens, and Ones
EMC 3072
© Evan-Moor Corp.

Hundreds, Tens, and Ones
EMC 3072
© Evan-Moor Corp.

Hundreds, Tens, and Ones
EMC 3072
© Evan-Moor Corp.

Hundreds, Tens, and Ones
EMC 3072
© Evan-Moor Corp.

Take It to Your Seat Centers

Greater, Less, or Equal?

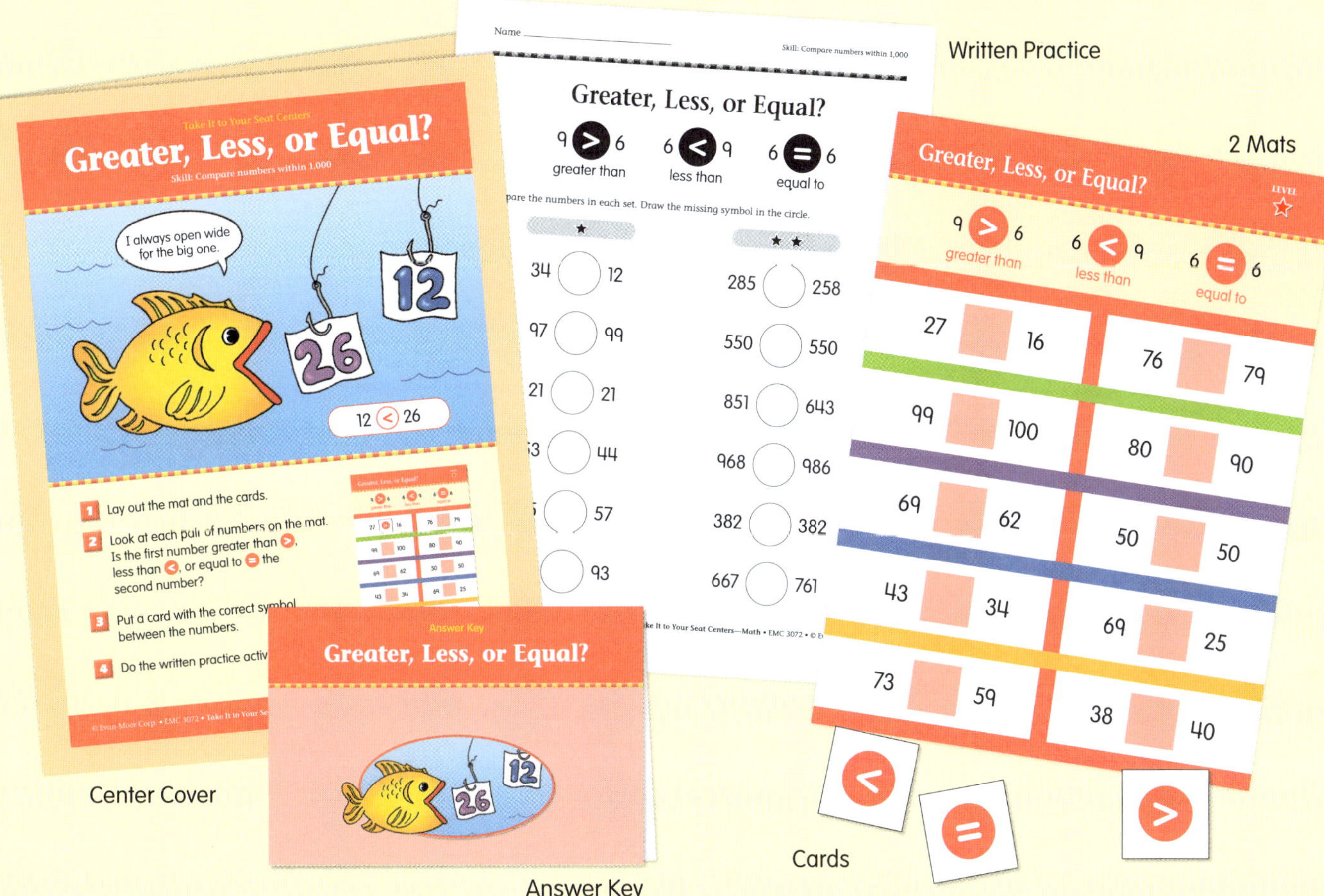

Written Practice

2 Mats

Center Cover

Answer Key

Cards

Skill: Compare numbers within 1,000, using greater than, less than, and equal to symbols

Steps to Follow

1. **Prepare the center.** (See page 3.)
2. **Introduce the center.** State the goal. Say: *You will place a symbol card for greater than, less than, or equal to between each pair of numbers on the mat to show how the numbers are related.*
3. **Teach the skill.** Demonstrate how to use the center with individual students or small groups.
4. **Practice the skill.** Have students use the center independently or with a partner.

Contents

Written Practice 34

Center Cover 35

Answer Key 37

Center Mats

Level 1 39

Level 2 41

Cards 43

Name ____________________ Skill: Compare numbers within 1,000

Greater, Less, or Equal?

$9 > 6$ greater than

$6 < 9$ less than

$6 = 6$ equal to

Compare the numbers in each set. Draw the missing symbol in the circle.

★

34 ◯ 12

97 ◯ 99

21 ◯ 21

53 ◯ 44

65 ◯ 57

82 ◯ 93

★ ★

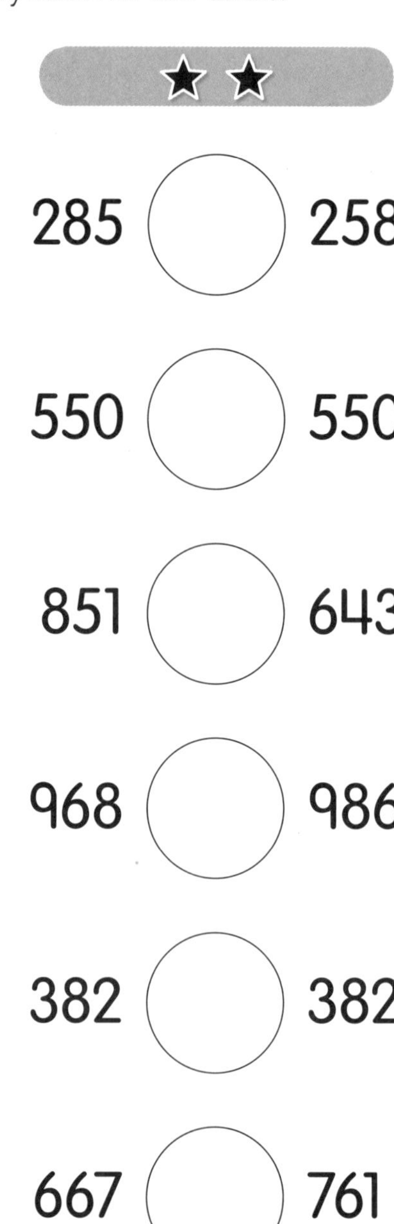

285 ◯ 258

550 ◯ 550

851 ◯ 643

968 ◯ 986

382 ◯ 382

667 ◯ 761

Take It to Your Seat Centers

Greater, Less, or Equal?

Skill: Compare numbers within 1,000

I always open wide for the big one.

26

12

12 < 26

1. Lay out the mat and the cards.
2. Look at each pair of numbers on the mat. Is the first number greater than >, less than <, or equal to = the second number?
3. Put a card with the correct symbol between the numbers.
4. Do the written practice activity.

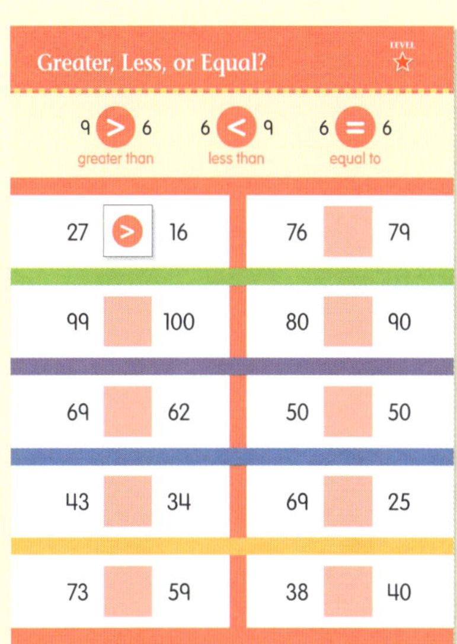

Answer Key

Greater, Less, or Equal?

(fold)

Written Practice

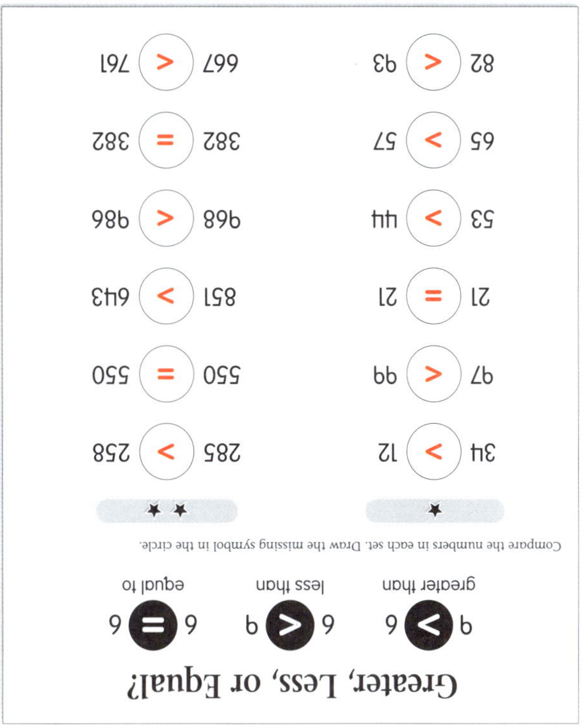

Greater, Less, or Equal?

9 > 6 greater than

6 < 9 less than

6 = 6 equal to

Compare the numbers in each set. Draw the missing symbol in the circle.

★	★ ★
34 > 12	285 > 258
97 < 99	550 = 550
21 = 21	851 > 643
53 > 44	968 < 986
65 > 57	382 = 382
82 < 93	667 < 761

Answer Key

Greater, Less, or Equal?

Greater, Less, or Equal?

LEVEL ☆

9 > 6 greater than | 6 < 9 less than | 6 = 6 equal to

27	>	16	76	<	79
99	<	100	80	<	90
69	>	62	50	=	50
43	>	34	69	>	25
73	>	59	38	<	40

Greater, Less, or Equal?

LEVEL ☆☆

9 > 6 greater than | 6 < 9 less than | 6 = 6 equal to

123	<	319	600	<	800
460	>	243	299	>	120
900	=	900	315	<	911
765	>	567	952	>	529
528	<	582	699	<	1,000

Greater, Less, or Equal?

LEVEL 

9 > 6

greater than

6 < 9

less than

6 = 6

equal to

27		16	76		79
99		100	80		90
69		62	50		50
43		34	69		25
73		59	38 		40

Greater, Less, or Equal?

LEVEL

9 > 6
greater than

6 < 9
less than

6 = 6
equal to

123 ☐ 319

600  800

460 ☐ 243

299  120

900 ☐ 900

315  911

765  567

952  529

528  582

699  1,000

<	<	<	<	<	<	<
<	<	<	<	<	<	<
<	<	<	<	<	<	<
=	=	=	=	=	=	=
<	<	<	<	<	<	<
<	<	<	<	<	<	<
<	<	<	<	<	<	<
=	=	=	=	=	=	=

Greater, Less, or Equal?
EMC 3072
© Evan-Moor Corp.

Greater, Less, or Equal?
EMC 3072
© Evan-Moor Corp.

Greater, Less, or Equal?
EMC 3072
© Evan-Moor Corp.

Greater, Less, or Equal?
EMC 3072
© Evan-Moor Corp.

Greater, Less, or Equal?
EMC 3072
© Evan-Moor Corp.

Greater, Less, or Equal?
EMC 3072
© Evan-Moor Corp.

Greater, Less, or Equal?
EMC 3072
© Evan-Moor Corp.

Greater, Less, or Equal?
EMC 3072
© Evan-Moor Corp.

Greater, Less, or Equal?
EMC 3072
© Evan-Moor Corp.

Greater, Less, or Equal?
EMC 3072
© Evan-Moor Corp.

Greater, Less, or Equal?
EMC 3072
© Evan-Moor Corp.

Greater, Less, or Equal?
EMC 3072
© Evan-Moor Corp.

Greater, Less, or Equal?
EMC 3072
© Evan-Moor Corp.

Greater, Less, or Equal?
EMC 3072
© Evan-Moor Corp.

Greater, Less, or Equal?
EMC 3072
© Evan-Moor Corp.

Greater, Less, or Equal?
EMC 3072
© Evan-Moor Corp.

Greater, Less, or Equal?
EMC 3072
© Evan-Moor Corp.

Greater, Less, or Equal?
EMC 3072
© Evan-Moor Corp.

Greater, Less, or Equal?
EMC 3072
© Evan-Moor Corp.

Greater, Less, or Equal?
EMC 3072
© Evan-Moor Corp.

Greater, Less, or Equal?
EMC 3072
© Evan-Moor Corp.

Greater, Less, or Equal?
EMC 3072
© Evan-Moor Corp.

Greater, Less, or Equal?
EMC 3072
© Evan-Moor Corp.

Greater, Less, or Equal?
EMC 3072
© Evan-Moor Corp.

Greater, Less, or Equal?
EMC 3072
© Evan-Moor Corp.

Greater, Less, or Equal?
EMC 3072
© Evan-Moor Corp.

Greater, Less, or Equal?
EMC 3072
© Evan-Moor Corp.

Greater, Less, or Equal?
EMC 3072
© Evan-Moor Corp.

Greater, Less, or Equal?
EMC 3072
© Evan-Moor Corp.

Greater, Less, or Equal?
EMC 3072
© Evan-Moor Corp.

Greater, Less, or Equal?
EMC 3072
© Evan-Moor Corp.

Greater, Less, or Equal?
EMC 3072
© Evan-Moor Corp.

Greater, Less, or Equal?
EMC 3072
© Evan-Moor Corp.

Greater, Less, or Equal?
EMC 3072
© Evan-Moor Corp.

Greater, Less, or Equal?
EMC 3072
© Evan-Moor Corp.

Greater, Less, or Equal?
EMC 3072
© Evan-Moor Corp.

Greater, Less, or Equal?
EMC 3072
© Evan-Moor Corp.

Greater, Less, or Equal?
EMC 3072
© Evan-Moor Corp.

Greater, Less, or Equal?
EMC 3072
© Evan-Moor Corp.

Greater, Less, or Equal?
EMC 3072
© Evan-Moor Corp.

Greater, Less, or Equal?
EMC 3072
© Evan-Moor Corp.

Greater, Less, or Equal?
EMC 3072
© Evan-Moor Corp.

Greater, Less, or Equal?
EMC 3072
© Evan-Moor Corp.

Greater, Less, or Equal?
EMC 3072
© Evan-Moor Corp.

Greater, Less, or Equal?
EMC 3072
© Evan-Moor Corp.

Greater, Less, or Equal?
EMC 3072
© Evan-Moor Corp.

Greater, Less, or Equal?
EMC 3072
© Evan-Moor Corp.

Greater, Less, or Equal?
EMC 3072
© Evan-Moor Corp.

Greater, Less, or Equal?
EMC 3072
© Evan-Moor Corp.

Greater, Less, or Equal?
EMC 3072
© Evan-Moor Corp.

Greater, Less, or Equal?
EMC 3072
© Evan-Moor Corp.

Greater, Less, or Equal?
EMC 3072
© Evan-Moor Corp.

Greater, Less, or Equal?
EMC 3072
© Evan-Moor Corp.

Greater, Less, or Equal?
EMC 3072
© Evan-Moor Corp.

Greater, Less, or Equal?
EMC 3072
© Evan-Moor Corp.

Greater, Less, or Equal?
EMC 3072
© Evan-Moor Corp.

Take It to Your Seat Centers

Skip Counting to 1,000

Skill: Count by 5s, 10s, and 100s

Steps to Follow

1. **Prepare the center.** (See page 3.)
2. **Introduce the center.** State the goal. Say: *You will place number cards on each mat in the correct order to count by 5s, 10s, and 100s.*
3. **Teach the skill.** Demonstrate how to use the center with individual students or small groups.
4. **Practice the skill.** Have students use the center independently or with a partner.

Contents

Written Practice............46
Center Cover..................47
Answer Key....................49
Center Mats............. 51, 53
Cards............................55

Name ______________________

Skill: Count by 5s, 10s, and 100s

Skip Counting to 1,000

Count by **5s** to fill in the missing numbers on the chart below.
Then color the correct numbers on the chart to count by **10s**.

5	10	15	20		30		40		
55		65		75			90		100
		115				135			
	160			175				195	
			220						250
		265					290		
305					330				
	360			375				395	
				425		435			
455			470						500

Write the missing numbers on the lines below to count by **100s** to 1,000.

100 ______ ______ ______ ______

______ ______ ______ ______ 1,000

Skip Counting to 1,000

Skill: Count by 5s, 10s, and 100s

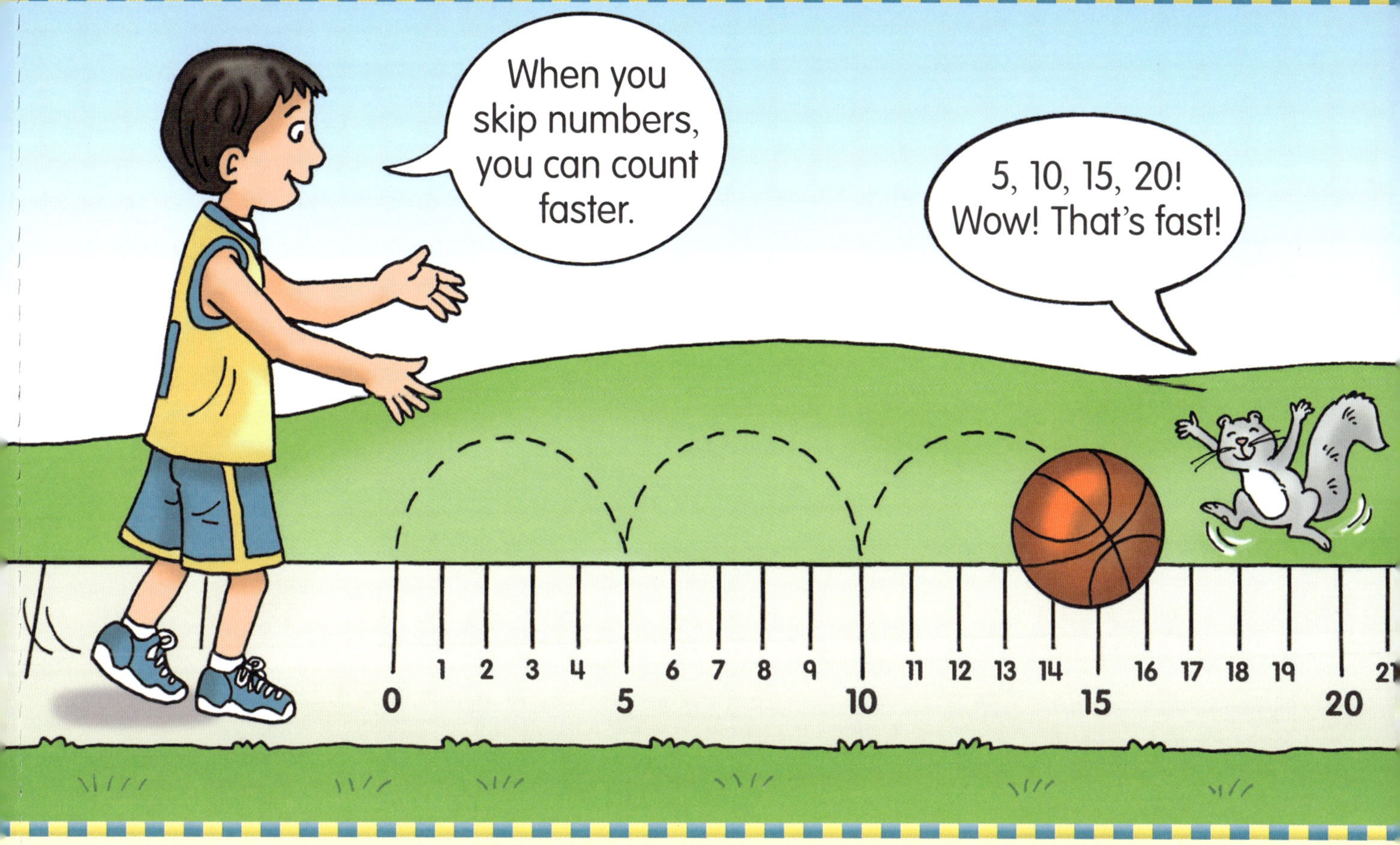

When you skip numbers, you can count faster.
5, 10, 15, 20! Wow! That's fast!
0 1 2 3 4 5 6 7 8 9 10 11 12 13 14 15 16 17 18 19 20

1. Lay out the mats and the number cards.
2. Put the correct number card in each empty square on the mats to count by 5s, 10s, or 100s. Count as you go to see the pattern that the numbers make.
3. Do the written practice activity.

Skip Counting to 1,000
Count by 10s
Count by 100s
Skip Counting to 1,000
Count by 5s

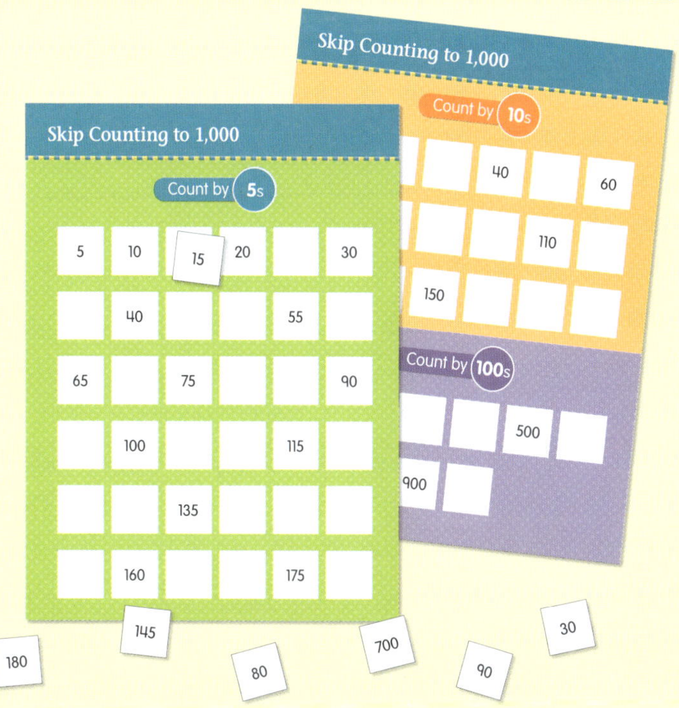

Written Practice

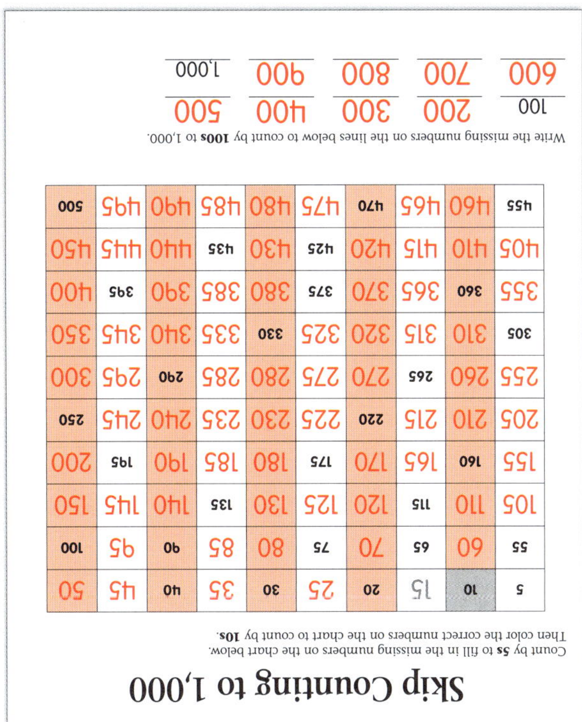

Skip Counting to 1,000

Count by **5s** to fill in the missing numbers on the chart below.
Then color the correct numbers on the chart to count by **10s**.

5	10	15	20	25	30	35	40	45	50
55	60	65	70	75	80	85	90	95	100
105	110	115	120	125	130	135	140	145	150
155	160	165	170	175	180	185	190	195	200
205	210	215	220	225	230	235	240	245	250
255	260	265	270	275	280	285	290	295	300
305	310	315	320	325	330	335	340	345	350
355	360	365	370	375	380	385	390	395	400
405	410	415	420	425	430	435	440	445	450
455	460	465	470	475	480	485	490	495	500

Write the missing numbers on the lines below to count by **100s** to 1,000.

100 200 300 400 500
600 700 800 900 1,000

(fold)

Answer Key

Skip Counting to 1,000

Answer Key

Skip Counting to 1,000

Skip Counting to 1,000

Count by 5s

5	10	15	20	25	30
35	40	45	50	55	60
65	70	75	80	85	90
95	100	105	110	115	120
125	130	135	140	145	150
155	160	165	170	175	180

Skip Counting to 1,000

Count by 10s

10	20	30	40	50	60
70	80	90	100	110	120
130	140	150	160	170	180

Count by 100s

100	200	300	400	500	600
700	800	900	1,000		

Skip Counting to 1,000

Count by 5s

5	10		20		30
	40			55	
65		75			90
	100			115	
		135			
	160			175	

Take It to Your Seat Centers—Math • EMC 3072 • © Evan-Moor Corp.

Skip Counting to 1,000

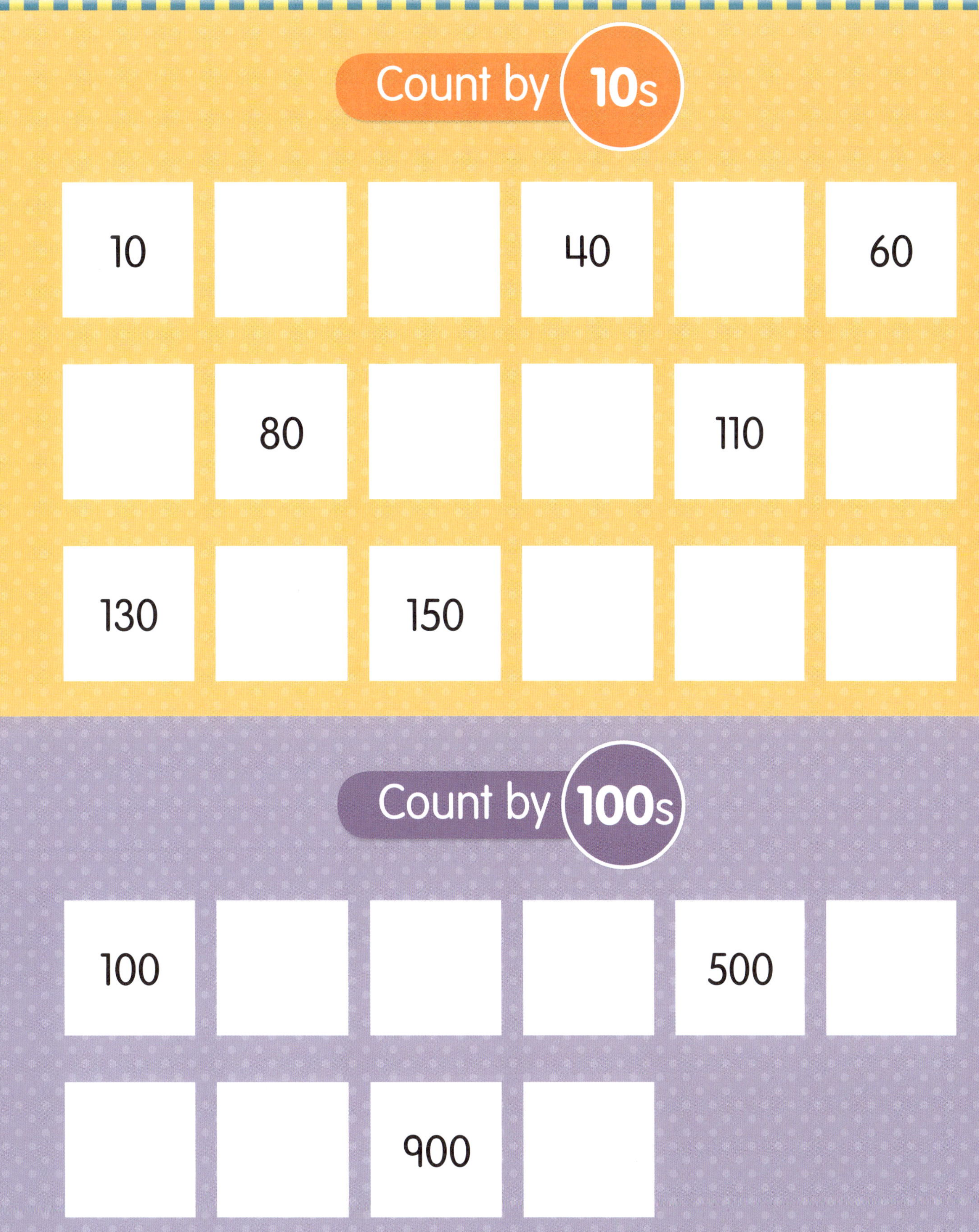

15	25	35	45	50
60	70	80	85	95
105	110	120	125	130
140	145	150	155	165
170	180	20	30	50
70	90	100	120	140
160	170	180	200	300
400	600	700	800	1,000

Skip Counting to 1,000 EMC 3072 © Evan-Moor Corp.	Skip Counting to 1,000 EMC 3072 © Evan-Moor Corp.	Skip Counting to 1,000 EMC 3072 © Evan-Moor Corp.	Skip Counting to 1,000 EMC 3072 © Evan-Moor Corp.	Skip Counting to 1,000 EMC 3072 © Evan-Moor Corp.
Skip Counting to 1,000 EMC 3072 © Evan-Moor Corp.	Skip Counting to 1,000 EMC 3072 © Evan-Moor Corp.	Skip Counting to 1,000 EMC 3072 © Evan-Moor Corp.	Skip Counting to 1,000 EMC 3072 © Evan-Moor Corp.	Skip Counting to 1,000 EMC 3072 © Evan-Moor Corp.
Skip Counting to 1,000 EMC 3072 © Evan-Moor Corp.	Skip Counting to 1,000 EMC 3072 © Evan-Moor Corp.	Skip Counting to 1,000 EMC 3072 © Evan-Moor Corp.	Skip Counting to 1,000 EMC 3072 © Evan-Moor Corp.	Skip Counting to 1,000 EMC 3072 © Evan-Moor Corp.
Skip Counting to 1,000 EMC 3072 © Evan-Moor Corp.	Skip Counting to 1,000 EMC 3072 © Evan-Moor Corp.	Skip Counting to 1,000 EMC 3072 © Evan-Moor Corp.	Skip Counting to 1,000 EMC 3072 © Evan-Moor Corp.	Skip Counting to 1,000 EMC 3072 © Evan-Moor Corp.
Skip Counting to 1,000 EMC 3072 © Evan-Moor Corp.	Skip Counting to 1,000 EMC 3072 © Evan-Moor Corp.	Skip Counting to 1,000 EMC 3072 © Evan-Moor Corp.	Skip Counting to 1,000 EMC 3072 © Evan-Moor Corp.	Skip Counting to 1,000 EMC 3072 © Evan-Moor Corp.
Skip Counting to 1,000 EMC 3072 © Evan-Moor Corp.	Skip Counting to 1,000 EMC 3072 © Evan-Moor Corp.	Skip Counting to 1,000 EMC 3072 © Evan-Moor Corp.	Skip Counting to 1,000 EMC 3072 © Evan-Moor Corp.	Skip Counting to 1,000 EMC 3072 © Evan-Moor Corp.
Skip Counting to 1,000 EMC 3072 © Evan-Moor Corp.	Skip Counting to 1,000 EMC 3072 © Evan-Moor Corp.	Skip Counting to 1,000 EMC 3072 © Evan-Moor Corp.	Skip Counting to 1,000 EMC 3072 © Evan-Moor Corp.	Skip Counting to 1,000 EMC 3072 © Evan-Moor Corp.
Skip Counting to 1,000 EMC 3072 © Evan-Moor Corp.	Skip Counting to 1,000 EMC 3072 © Evan-Moor Corp.	Skip Counting to 1,000 EMC 3072 © Evan-Moor Corp.	Skip Counting to 1,000 EMC 3072 © Evan-Moor Corp.	Skip Counting to 1,000 EMC 3072 © Evan-Moor Corp.

Take It to Your Seat Centers

Facts to 20

Skill: Build fluency with addition and subtraction facts to 20

Steps to Follow

1. **Prepare the center.** (See page 3.)
2. **Introduce the center.** State the goal. Say: *You will place each dog card in the doghouse showing the number that completes the equation.*
3. **Teach the skill.** Demonstrate how to use the center with individual students or small groups.
4. **Practice the skill.** Have students use the center independently or with a partner.

Contents

Written Practice............ 58

Center Cover.................. 59

Answer Key.................... 61

Center Mats 63, 65, 67

Cards 69

Facts to 20

Complete each equation below. Then look at the key and write the letter for each answer on the line below the bone.

Key

4 = **e**	5 = **d**	6 = **k**
7 = **l**	8 = **o**	9 = **i**
10 = **s**	11 = **f**	12 = **g**

4 + 1 = 5	5 + 3 =	17 − 5 =	20 − 10 =
d	___	___	___
3 + 4 =	2 + 7 =	14 − 8 =	6 − 2 =
___	___	___	___
7 + 4 =	10 − 2 =	15 − 7 =	13 − 8 =
___	___	___	___

Write the words that the letters spell as a sentence.

____

Facts to 20

Skill: Build fluency with addition and subtraction facts to 20

10 – 6

2 + 2

3 + 1

6 – 2

4

1. Lay out the mats and the cards.
2. Look at the dog on each card and add or subtract the numbers.
3. Put the card in the doghouse that shows the answer.
4. Do the written practice activity.

Written Practice

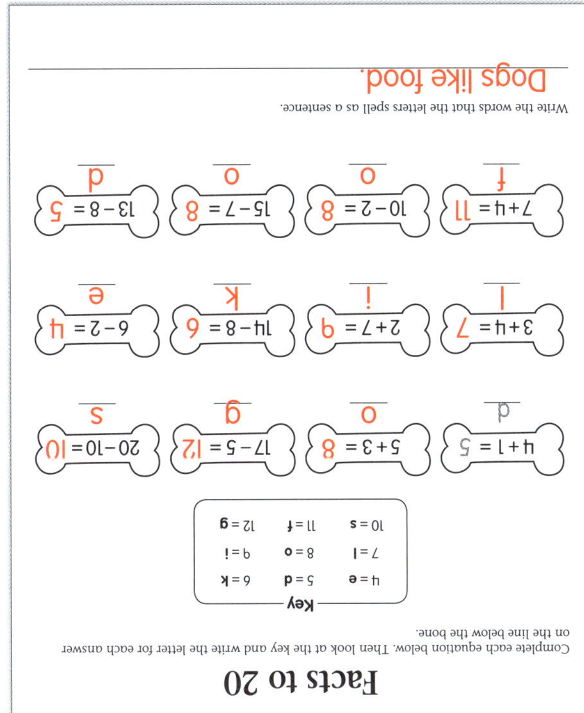

Facts to 20

Complete each equation below. Then look at the key and write the letter for each answer on the line below the bone.

Key

4 = e	5 = d	6 = k
7 = l	8 = o	9 = i
10 = s	11 = f	12 = g

4 + 1 = 5	5 + 3 = 8	17 − 5 = 12	20 − 10 = 10
d	o	g	s
3 + 4 = 7	2 + 7 = 9	14 − 8 = 6	6 − 2 = 4
l	i	k	e
7 + 4 = 11	10 − 2 = 8	15 − 7 = 8	13 − 8 = 5
f	o	o	d

Write the words that the letters spell as a sentence.

Dogs like food.

(fold)

Answer Key

Facts to 20

Answer Key

Facts to 20

Facts to 20

Put the dogs in their houses as fast as you can.

Take It to Your Seat Centers—Math • EMC 3072 • © Evan-Moor Corp.

Facts to 20

7

8

9

Facts to 20

10

11

12

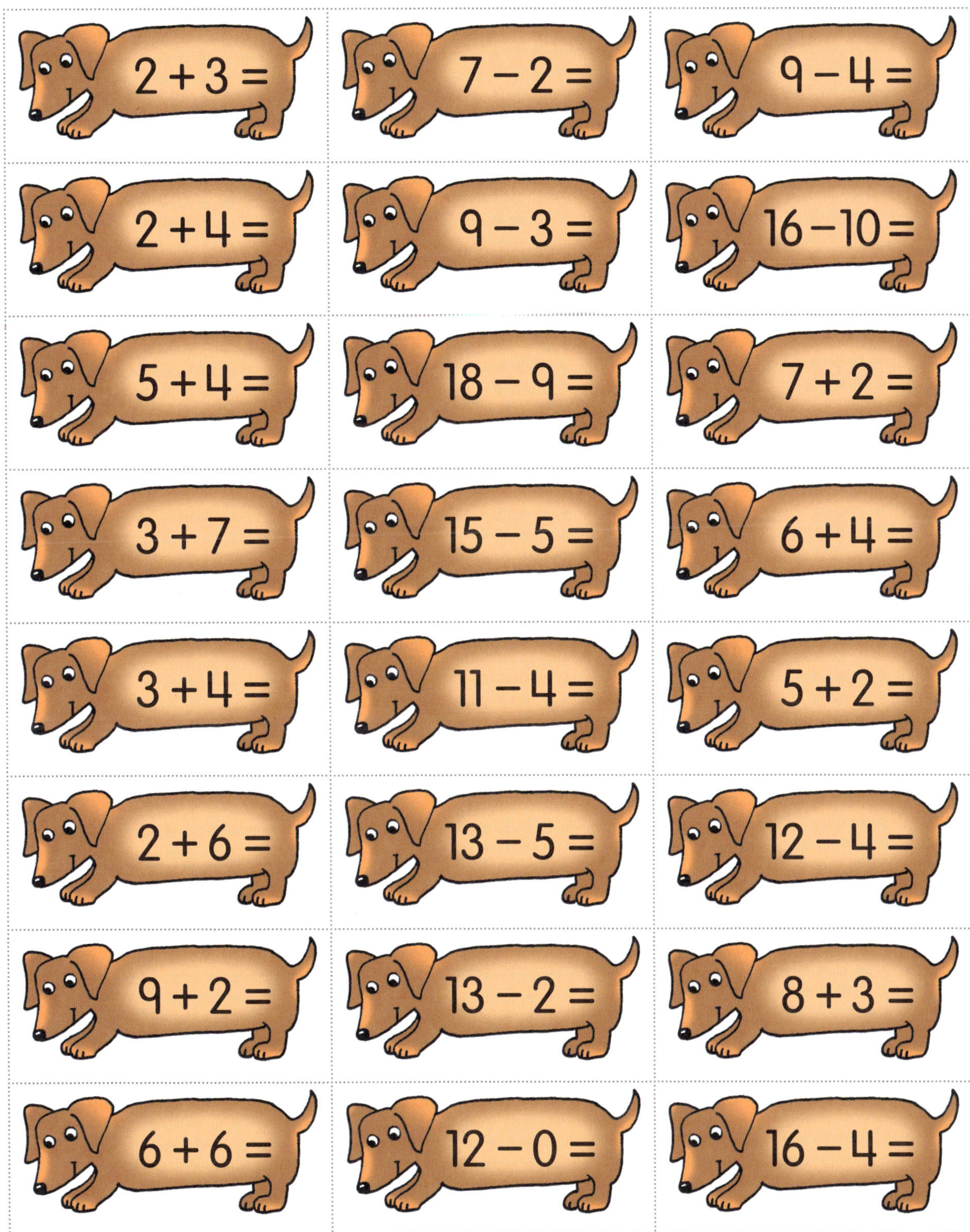
2 + 3 =
7 − 2 =
9 − 4 =
2 + 4 =
9 − 3 =
16 − 10 =
5 + 4 =
18 − 9 =
7 + 2 =
3 + 7 =
15 − 5 =
6 + 4 =
3 + 4 =
11 − 4 =
5 + 2 =
2 + 6 =
13 − 5 =
12 − 4 =
9 + 2 =
13 − 2 =
8 + 3 =
6 + 6 =
12 − 0 =
16 − 4 =

Facts to 20
EMC 3072
© Evan-Moor Corp.

Facts to 20
EMC 3072
© Evan-Moor Corp.

Facts to 20
EMC 3072
© Evan-Moor Corp.

Facts to 20
EMC 3072
© Evan-Moor Corp.

Facts to 20
EMC 3072
© Evan-Moor Corp.

Facts to 20
EMC 3072
© Evan-Moor Corp.

Facts to 20
EMC 3072
© Evan-Moor Corp.

Facts to 20
EMC 3072
© Evan-Moor Corp.

Facts to 20
EMC 3072
© Evan-Moor Corp.

Facts to 20
EMC 3072
© Evan-Moor Corp.

Facts to 20
EMC 3072
© Evan-Moor Corp.

Facts to 20
EMC 3072
© Evan-Moor Corp.

Facts to 20
EMC 3072
© Evan-Moor Corp.

Facts to 20
EMC 3072
© Evan-Moor Corp.

Facts to 20
EMC 3072
© Evan-Moor Corp.

Facts to 20
EMC 3072
© Evan-Moor Corp.

Facts to 20
EMC 3072
© Evan-Moor Corp.

Facts to 20
EMC 3072
© Evan-Moor Corp.

Facts to 20
EMC 3072
© Evan-Moor Corp.

Facts to 20
EMC 3072
© Evan-Moor Corp.

Facts to 20
EMC 3072
© Evan-Moor Corp.

Facts to 20
EMC 3072
© Evan-Moor Corp.

Facts to 20
EMC 3072
© Evan-Moor Corp.

Facts to 20
EMC 3072
© Evan-Moor Corp.

Take It to Your Seat Centers

Fact Families

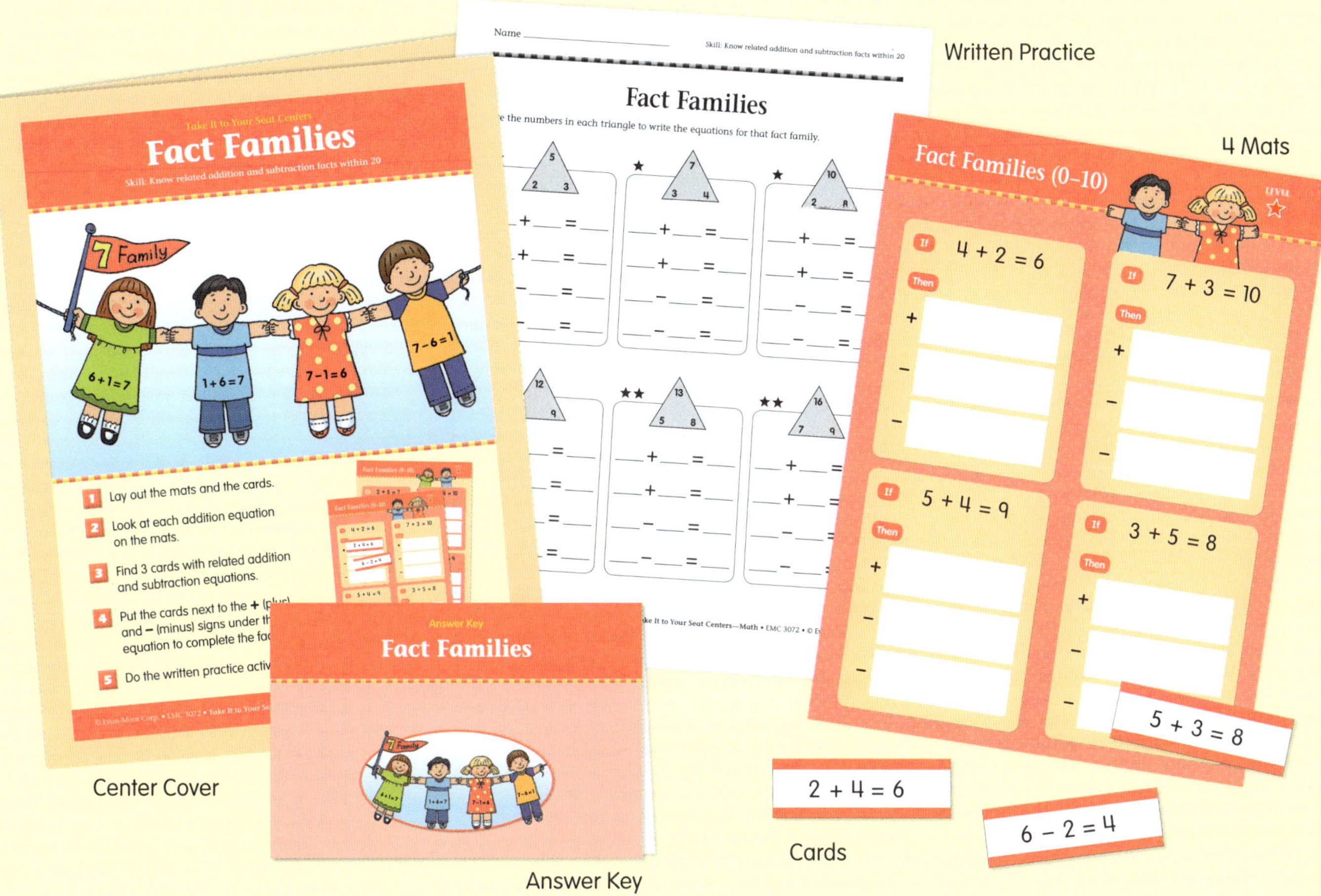

Skill: Know related addition and subtraction facts within 20

Steps to Follow

1. **Prepare the center.** (See page 3.)
2. **Introduce the center.** State the goal. Say: *You will find cards with the correct addition and subtraction equations to complete each fact family on the mats.*
3. **Teach the skill.** Demonstrate how to use the center with individual students or small groups.
4. **Practice the skill.** Have students use the center independently or with a partner.

Contents

Written Practice............. 72
Center Cover.................. 73
Answer Key..................... 75
Center Mats
Level 1.................. 77, 79
Level 2.................. 81, 83
Cards........................ 85, 87

Name ____________________ Skill: Know related addition and subtraction facts within 20

Fact Families

Use the numbers in each triangle to write the equations for that fact family.

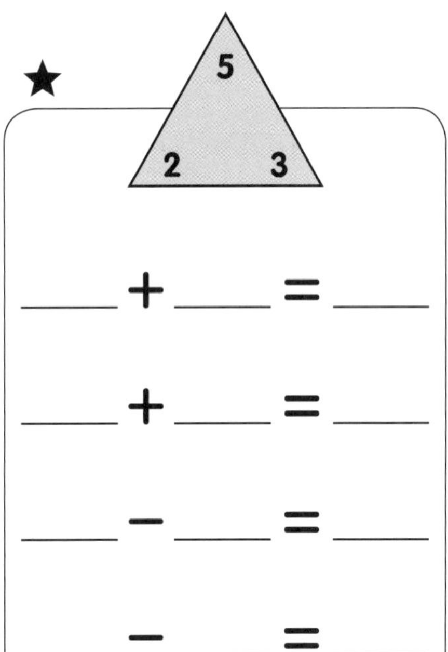

★

7

3 4

____ + ____ = ____

____ + ____ = ____

____ − ____ = ____

____ − ____ = ____

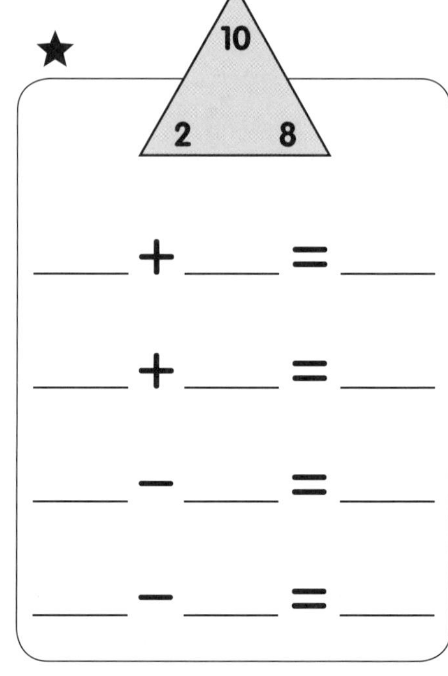

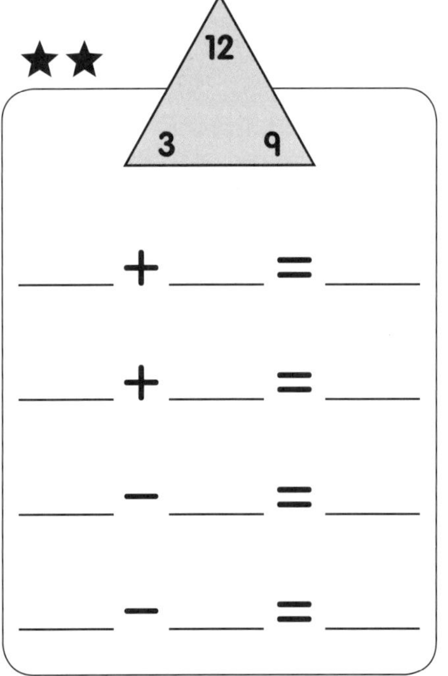

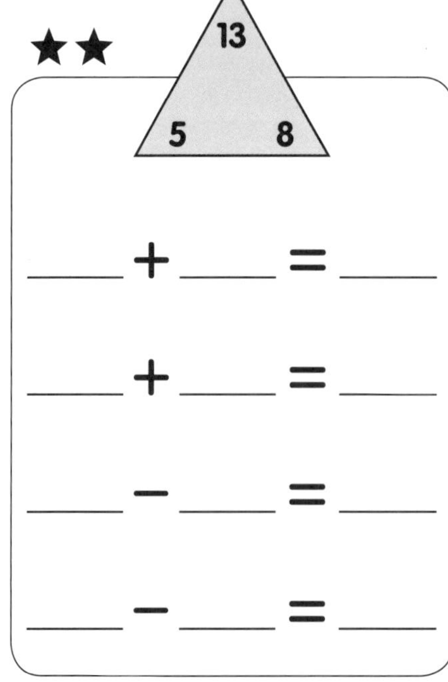

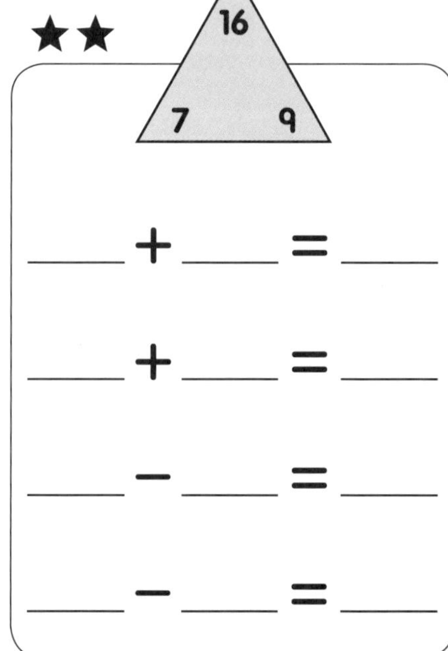

Take It to Your Seat Centers

Fact Families

Skill: Know related addition and subtraction facts within 20

7 Family

6 + 1 = 7

1 + 6 = 7

7 − 1 = 6

7 − 6 = 1

1. Lay out the mats and the cards.
2. Look at each addition equation on the mats.
3. Find 3 cards with related addition and subtraction equations.
4. Put the cards next to the **+** (plus) and **−** (minus) signs under the equation to complete the fact family.
5. Do the written practice activity.

Fact Families

Answer Key

(fold)

Written Practice

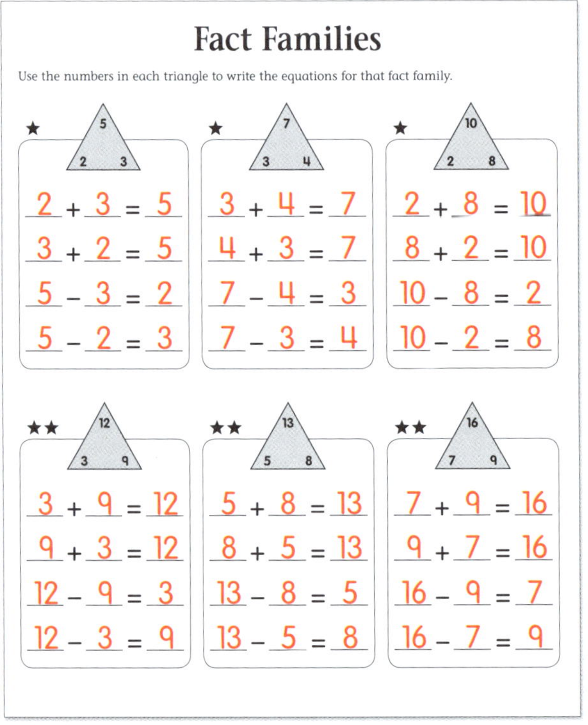

Fact Families

Use the numbers in each triangle to write the equations for that fact family.

★ (5; 2, 3)
2 + 3 = 5
3 + 2 = 5
5 − 3 = 2
5 − 2 = 3

★ (7; 3, 4)
3 + 4 = 7
4 + 3 = 7
7 − 4 = 3
7 − 3 = 4

★ (10; 2, 8)
2 + 8 = 10
8 + 2 = 10
10 − 8 = 2
10 − 2 = 8

★★ (12; 3, 9)
3 + 9 = 12
9 + 3 = 12
12 − 9 = 3
12 − 3 = 9

★★ (13; 5, 8)
5 + 8 = 13
8 + 5 = 13
13 − 8 = 5
13 − 5 = 8

★★ (16; 7, 9)
7 + 9 = 16
9 + 7 = 16
16 − 9 = 7
16 − 7 = 9

Answer Key

Fact Families

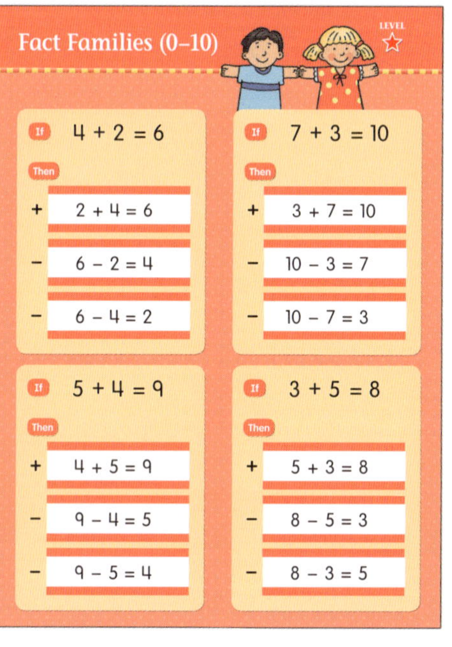

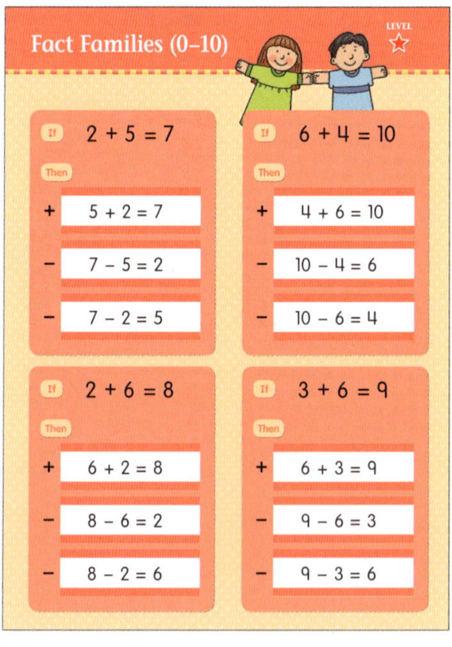

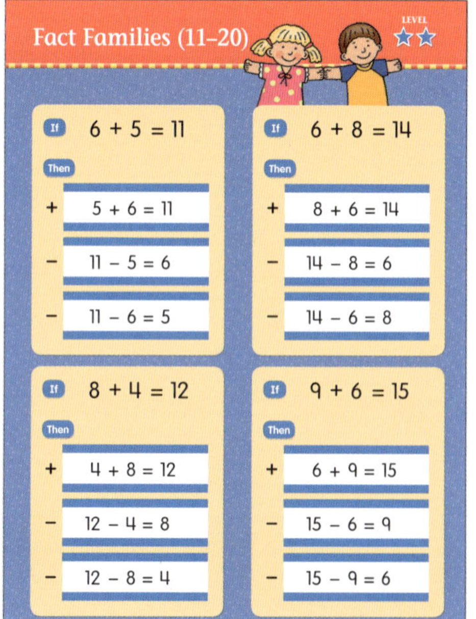

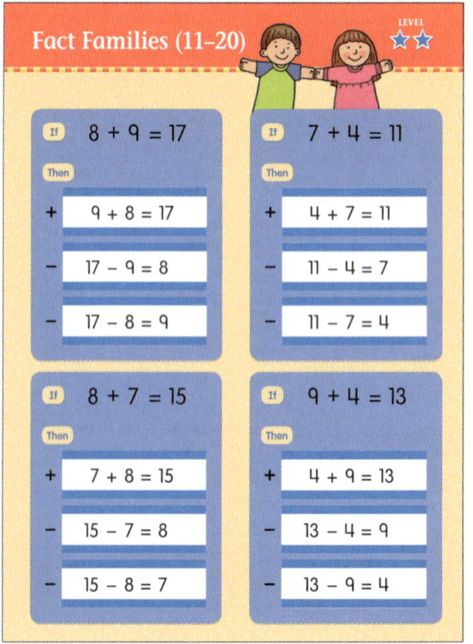

Fact Families (0–10)

If $4 + 2 = 6$

Then

+

−

−

If $7 + 3 = 10$

Then

+

−

−

If $5 + 4 = 9$

Then

+

−

−

If $3 + 5 = 8$

Then

+

−

−

Fact Families (0–10)

LEVEL 

If $2 + 5 = 7$

Then

+

−

−

If $6 + 4 = 10$

Then

+

−

−

If $2 + 6 = 8$

Then

+

−

−

If $3 + 6 = 9$

Then

+

−

−

Fact Families (11–20)

If $6 + 5 = 11$

Then

+

−

−

If $6 + 8 = 14$

Then

+

−

−

If $8 + 4 = 12$

Then

+

−

−

If $9 + 6 = 15$

Then

+

−

−

Fact Families (11–20)

LEVEL ★★

If $8 + 9 = 17$

Then

+

−

−

If $7 + 4 = 11$

Then

+

−

−

If $8 + 7 = 15$

Then

+

−

−

If $9 + 4 = 13$

Then

+

−

−

2 + 4 = 6	6 − 2 = 4	6 − 4 = 2
3 + 7 = 10	10 − 3 = 7	10 − 7 = 3
4 + 5 = 9	9 − 4 = 5	9 − 5 = 4
5 + 3 = 8	8 − 5 = 3	8 − 3 = 5
5 + 2 = 7	7 − 5 = 2	7 − 2 = 5
4 + 6 = 10	10 − 4 = 6	10 − 6 = 4
6 + 2 = 8	8 − 6 = 2	8 − 2 = 6
6 + 3 = 9	9 − 6 = 3	9 − 3 = 6

Fact Families
EMC 3072
© Evan-Moor Corp.

Fact Families
EMC 3072
© Evan-Moor Corp.

Fact Families
EMC 3072
© Evan-Moor Corp.

Fact Families
EMC 3072
© Evan-Moor Corp.

Fact Families
EMC 3072
© Evan-Moor Corp.

Fact Families
EMC 3072
© Evan-Moor Corp.

Fact Families
EMC 3072
© Evan-Moor Corp.

Fact Families
EMC 3072
© Evan-Moor Corp.

Fact Families
EMC 3072
© Evan-Moor Corp.

Fact Families
EMC 3072
© Evan-Moor Corp.

Fact Families
EMC 3072
© Evan-Moor Corp.

Fact Families
EMC 3072
© Evan-Moor Corp.

Fact Families
EMC 3072
© Evan-Moor Corp.

Fact Families
EMC 3072
© Evan-Moor Corp.

Fact Families
EMC 3072
© Evan-Moor Corp.

Fact Families
EMC 3072
© Evan-Moor Corp.

Fact Families
EMC 3072
© Evan-Moor Corp.

Fact Families
EMC 3072
© Evan-Moor Corp.

Fact Families
EMC 3072
© Evan-Moor Corp.

Fact Families
EMC 3072
© Evan-Moor Corp.

Fact Families
EMC 3072
© Evan-Moor Corp.

Fact Families
EMC 3072
© Evan-Moor Corp.

Fact Families
EMC 3072
© Evan-Moor Corp.

Fact Families
EMC 3072
© Evan-Moor Corp.

$5 + 6 = 11$	$11 - 5 = 6$	$11 - 6 = 5$
$8 + 6 = 14$	$14 - 8 = 6$	$14 - 6 = 8$
$4 + 8 = 12$	$12 - 4 = 8$	$12 - 8 = 4$
$6 + 9 = 15$	$15 - 6 = 9$	$15 - 9 = 6$
$9 + 8 = 17$	$17 - 9 = 8$	$17 - 8 = 9$
$4 + 7 = 11$	$11 - 4 = 7$	$11 - 7 = 4$
$7 + 8 = 15$	$15 - 7 = 8$	$15 - 8 = 7$
$4 + 9 = 13$	$13 - 9 = 4$	$13 - 4 = 9$

Fact Families EMC 3072 © Evan-Moor Corp.	**Fact Families** EMC 3072 © Evan-Moor Corp.	**Fact Families** EMC 3072 © Evan-Moor Corp.
Fact Families EMC 3072 © Evan-Moor Corp.	**Fact Families** EMC 3072 © Evan-Moor Corp.	**Fact Families** EMC 3072 © Evan-Moor Corp.
Fact Families EMC 3072 © Evan-Moor Corp.	**Fact Families** EMC 3072 © Evan-Moor Corp.	**Fact Families** EMC 3072 © Evan-Moor Corp.
Fact Families EMC 3072 © Evan-Moor Corp.	**Fact Families** EMC 3072 © Evan-Moor Corp.	**Fact Families** EMC 3072 © Evan-Moor Corp.
Fact Families EMC 3072 © Evan-Moor Corp.	**Fact Families** EMC 3072 © Evan-Moor Corp.	**Fact Families** EMC 3072 © Evan-Moor Corp.
Fact Families EMC 3072 © Evan-Moor Corp.	**Fact Families** EMC 3072 © Evan-Moor Corp.	**Fact Families** EMC 3072 © Evan-Moor Corp.
Fact Families EMC 3072 © Evan-Moor Corp.	**Fact Families** EMC 3072 © Evan-Moor Corp.	**Fact Families** EMC 3072 © Evan-Moor Corp.
Fact Families EMC 3072 © Evan-Moor Corp.	**Fact Families** EMC 3072 © Evan-Moor Corp.	**Fact Families** EMC 3072 © Evan-Moor Corp.

Take It to Your Seat Centers

Start at the Ones Place

Skill: Use place value to add and subtract

Steps to Follow

1. **Prepare the center.** (See page 3.)
2. **Introduce the center.** State the goal. Say: *You will start at the ones place to complete the addition or subtraction equation on each card.*
3. **Teach the skill.** Demonstrate how to use the center with individual students or small groups.
4. **Practice the skill.** Have students use the center independently or with a partner.

Contents

Response Form 90
Center Cover 91
Answer Key 93
Center Mat 95
Cards 97, 99

Name ______________________ Skill: Use place value to add and subtract

Start at the Ones Place

Write the answer to the equation on each card under the same equation below.

1. $\begin{array}{r} 62 \\ +36 \\ \hline \end{array}$

2. $\begin{array}{r} 79 \\ -15 \\ \hline \end{array}$

3. $\begin{array}{r} 63 \\ -42 \\ \hline \end{array}$

4. $\begin{array}{r} 57 \\ +30 \\ \hline \end{array}$

5. $\begin{array}{r} 66 \\ +13 \\ \hline \end{array}$

6. $\begin{array}{r} 87 \\ -24 \\ \hline \end{array}$

7. $\begin{array}{r} 41 \\ +52 \\ \hline \end{array}$

8. $\begin{array}{r} 99 \\ -67 \\ \hline \end{array}$

9. $\begin{array}{r} 557 \\ +401 \\ \hline \end{array}$

10. $\begin{array}{r} 387 \\ -262 \\ \hline \end{array}$

11. $\begin{array}{r} 236 \\ +453 \\ \hline \end{array}$

12. $\begin{array}{r} 496 \\ -234 \\ \hline \end{array}$

13. $\begin{array}{r} 625 \\ -322 \\ \hline \end{array}$

14. $\begin{array}{r} 412 \\ +303 \\ \hline \end{array}$

15. $\begin{array}{r} 999 \\ -535 \\ \hline \end{array}$

16. $\begin{array}{r} 235 \\ +342 \\ \hline \end{array}$

Take It to Your Seat Centers

Start at the Ones Place

Skill: Use place value to add and subtract

1. Lay out the mat.
2. Sort the cards into two groups: equation cards and number cards. Stack the equation cards in number order.
3. Put an equation card in the box on the mat and do the addition or subtraction. Remember to start at the ones place.
4. Put number cards in the squares under the equation to show the answer. Then write the answer under the same equation on the response form.
5. Repeat steps 3 and 4 with each equation card.

Answer Key

Start at the Ones Place

(fold)

Response Form

Start at the Ones Place

Write the answer to the equation on each card under the same equation below.

1. 62 + 36 = 98	2. 79 − 15 = 64	3. 63 − 42 = 21	4. 57 + 30 = 87
5. 66 + 13 = 79	6. 87 − 24 = 63	7. 41 + 52 = 93	8. 99 − 67 = 32
9. 557 + 401 = 958	10. 387 − 262 = 125	11. 236 + 453 = 689	12. 496 − 234 = 262
13. 625 − 322 = 303	14. 412 + 303 = 715	15. 999 − 535 = 464	16. 235 + 342 = 577

Answer Key

Start at the Ones Place

1. 62 + 36 = 98

2. 79 − 15 = 64

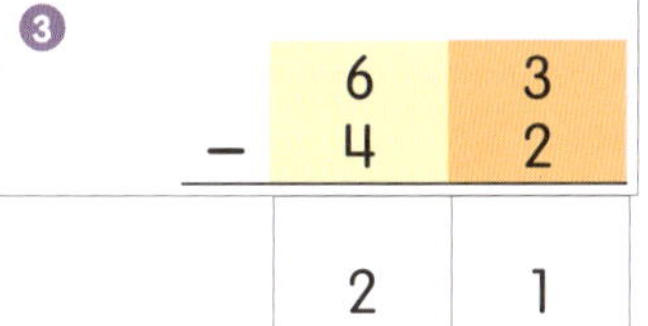

3. 63 − 42 = 21

4. 57 + 30 = 87

5. 66 + 13 = 79

6. 87 − 24 = 63

7. 41 + 52 = 93

8. 99 − 67 = 32

9. 557 + 401 = 958

10. 387 − 262 = 125

11. 236 + 453 = 689

12. 496 − 234 = 262

13. 625 − 322 = 303

14. 412 + 303 = 715

15. 999 − 535 = 464

16. 235 + 342 = 577

Start at the Ones Place

Take It to Your Seat Centers—Math • EMC 3072 • © Evan-Moor Corp.

1

  6 2
+ 3 6

5

  6 6
+ 1 3

2

  7 9
− 1 5

8

  9 9
− 6 7

9

  5 5 7
+ 4 0 1

16

  2 3 5
+ 3 4 2

10

  3 8 7
− 2 6 2

15

  9 9 9
− 5 3 5

7

  4 1
+ 5 2

4

  5 7
+ 3 0

6

  8 7
− 2 4

3

  6 3
− 4 2

11

  2 3 6
+ 4 5 3

14

  4 1 2
+ 3 0 3

12

  4 9 6
− 2 3 4

13

  6 2 5
− 3 2 2

Start at the Ones Place

EMC 3072
© Evan-Moor Corp.

Start at the Ones Place

EMC 3072
© Evan-Moor Corp.

Start at the Ones Place

EMC 3072
© Evan-Moor Corp.

Start at the Ones Place

EMC 3072
© Evan-Moor Corp.

Start at the Ones Place

EMC 3072
© Evan-Moor Corp.

Start at the Ones Place

EMC 3072
© Evan-Moor Corp.

Start at the Ones Place

EMC 3072
© Evan-Moor Corp.

Start at the Ones Place

EMC 3072
© Evan-Moor Corp.

Start at the Ones Place

EMC 3072
© Evan-Moor Corp.

Start at the Ones Place

EMC 3072
© Evan-Moor Corp.

Start at the Ones Place

EMC 3072
© Evan-Moor Corp.

Start at the Ones Place

EMC 3072
© Evan-Moor Corp.

Start at the Ones Place

EMC 3072
© Evan-Moor Corp.

Start at the Ones Place

EMC 3072
© Evan-Moor Corp.

Start at the Ones Place

EMC 3072
© Evan-Moor Corp.

Start at the Ones Place

EMC 3072
© Evan-Moor Corp.

1 2 3 4 5 6 7 8 9

1 2 3 4 5 6 7 8 9

1 2 3 4 5 6 7 8 9

1 2 3 4 5 6 7 8 9

1 2 3 4 5 6 7 8 9

1 2 3 4 5 6 7 8 9

0 0 0 0 0 0 0 0 0

Start at the Ones Place
EMC 3072
© Evan-Moor Corp.

Start at the Ones Place
EMC 3072
© Evan-Moor Corp.

Start at the Ones Place
EMC 3072
© Evan-Moor Corp.

Start at the Ones Place
EMC 3072
© Evan-Moor Corp.

Start at the Ones Place
EMC 3072
© Evan-Moor Corp.

Start at the Ones Place
EMC 3072
© Evan-Moor Corp.

Start at the Ones Place
EMC 3072
© Evan-Moor Corp.

Start at the Ones Place
EMC 3072
© Evan-Moor Corp.

Start at the Ones Place
EMC 3072
© Evan-Moor Corp.

Start at the Ones Place
EMC 3072
© Evan-Moor Corp.

Start at the Ones Place
EMC 3072
© Evan-Moor Corp.

Start at the Ones Place
EMC 3072
© Evan-Moor Corp.

Start at the Ones Place
EMC 3072
© Evan-Moor Corp.

Start at the Ones Place
EMC 3072
© Evan-Moor Corp.

Start at the Ones Place
EMC 3072
© Evan-Moor Corp.

Start at the Ones Place
EMC 3072
© Evan-Moor Corp.

Start at the Ones Place
EMC 3072
© Evan-Moor Corp.

Start at the Ones Place
EMC 3072
© Evan-Moor Corp.

Start at the Ones Place
EMC 3072
© Evan-Moor Corp.

Start at the Ones Place
EMC 3072
© Evan-Moor Corp.

Start at the Ones Place
EMC 3072
© Evan-Moor Corp.

Start at the Ones Place
EMC 3072
© Evan-Moor Corp.

Start at the Ones Place
EMC 3072
© Evan-Moor Corp.

Start at the Ones Place
EMC 3072
© Evan-Moor Corp.

Start at the Ones Place
EMC 3072
© Evan-Moor Corp.

Start at the Ones Place
EMC 3072
© Evan-Moor Corp.

Start at the Ones Place
EMC 3072
© Evan-Moor Corp.

Start at the Ones Place
EMC 3072
© Evan-Moor Corp.

Start at the Ones Place
EMC 3072
© Evan-Moor Corp.

Start at the Ones Place
EMC 3072
© Evan-Moor Corp.

Start at the Ones Place
EMC 3072
© Evan-Moor Corp.

Start at the Ones Place
EMC 3072
© Evan-Moor Corp.

Start at the Ones Place
EMC 3072
© Evan-Moor Corp.

Start at the Ones Place
EMC 3072
© Evan-Moor Corp.

Start at the Ones Place
EMC 3072
© Evan-Moor Corp.

Start at the Ones Place
EMC 3072
© Evan-Moor Corp.

Start at the Ones Place
EMC 3072
© Evan-Moor Corp.

Start at the Ones Place
EMC 3072
© Evan-Moor Corp.

Start at the Ones Place
EMC 3072
© Evan-Moor Corp.

Start at the Ones Place
EMC 3072
© Evan-Moor Corp.

Start at the Ones Place
EMC 3072
© Evan-Moor Corp.

Start at the Ones Place
EMC 3072
© Evan-Moor Corp.

Start at the Ones Place
EMC 3072
© Evan-Moor Corp.

Start at the Ones Place
EMC 3072
© Evan-Moor Corp.

Start at the Ones Place
EMC 3072
© Evan-Moor Corp.

Start at the Ones Place
EMC 3072
© Evan-Moor Corp.

Start at the Ones Place
EMC 3072
© Evan-Moor Corp.

Start at the Ones Place
EMC 3072
© Evan-Moor Corp.

Start at the Ones Place
EMC 3072
© Evan-Moor Corp.

Start at the Ones Place
EMC 3072
© Evan-Moor Corp.

Start at the Ones Place
EMC 3072
© Evan-Moor Corp.

Start at the Ones Place
EMC 3072
© Evan-Moor Corp.

Start at the Ones Place
EMC 3072
© Evan-Moor Corp.

Start at the Ones Place
EMC 3072
© Evan-Moor Corp.

Start at the Ones Place
EMC 3072
© Evan-Moor Corp.

Start at the Ones Place
EMC 3072
© Evan-Moor Corp.

Start at the Ones Place
EMC 3072
© Evan-Moor Corp.

Start at the Ones Place
EMC 3072
© Evan-Moor Corp.

Start at the Ones Place
EMC 3072
© Evan-Moor Corp.

Start at the Ones Place
EMC 3072
© Evan-Moor Corp.

Start at the Ones Place
EMC 3072
© Evan-Moor Corp.

Start at the Ones Place
EMC 3072
© Evan-Moor Corp.

Start at the Ones Place
EMC 3072
© Evan-Moor Corp.

Take It to Your Seat Centers

Use a Ruler

* This center requires 12-inch and 30-cm rulers.

Skill: Measure length in standard units (inches and centimeters)

Steps to Follow

1. **Prepare the center.** (See page 3.) Provide rulers that measure up to 12 inches and 30 centimeters.
2. **Introduce the center.** State the goal. Say: *You will use a ruler to measure the length of each line on the mats in inches or centimeters.*
3. **Teach the skill.** Demonstrate how to use the center with individual students or small groups.
4. **Practice the skill.** Have students use the center independently or with a partner.

Contents

Response Form 102
Center Cover 103
Answer Key 105
Center Mats 107, 109
Cards 111

Use a Ruler

Look at the mats. Write the number of inches and centimeters you measured for each line. Then answer the questions.

a	_______ inches	a	_______ centimeters
b	_______ inches	b	_______ centimeters
c	_______ inches	c	_______ centimeters
d	_______ inches	d	_______ centimeters
e	_______ inches	e	_______ centimeters
f	_______ inches	f	_______ centimeters
g	_______ inches	g	_______ centimeters

1. What is the sum of the measurements for **c** and **g**?

   _______ inches _______ centimeters

2. How much longer is **a** than **f**?

   _______ inches _______ centimeters

3. What is the difference between the measurements for the longest line and the shortest line?

   _______ inches _______ centimeters

4. Which line is as long as **d** and **e** put together?

   inches: _______ centimeters: _______

 © Evan-Moor Corp.

Take It to Your Seat Centers

Use a Ruler

Skill: Measure length in standard units

I can measure in **inches** or **centimeters**.

* This center requires 12-inch and 30-cm rulers.

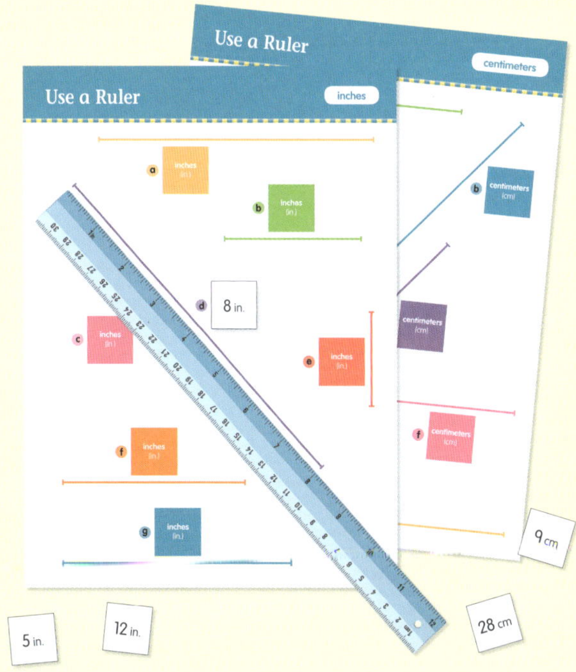

1. Lay out the mats and the cards.
2. Use a ruler to measure the lines on each mat in either **inches** or **centimeters**.
3. Put a card in the matching colored square to show how many inches or centimeters you measured for each line.
4. Complete the response form.

Response Form

Use a Ruler

Look at the mats. Write the number of inches and centimeters you measured for each line. Then answer the questions.

	Inches		Centimeters
a	6 inches	a	13 centimeters
b	3 inches	b	25 centimeters
c	10 inches	c	10 centimeters
d	8 inches	d	5 centimeters
e	2 inches	e	20 centimeters
f	4 inches	f	7 centimeters
g	5 inches	g	15 centimeters

1. What is the sum of the measurements for **c** and **g**?

   15 inches 25 centimeters

2. How much longer is **a** than **f**?

   2 inches 6 centimeters

3. What is the difference between the measurements for the longest line and the shortest line?

   8 inches 20 centimeters

4. Which line is as long as **d** and **e** put together?

   inches: c centimeters: b

(fold)

Answer Key

Use a Ruler

Answer Key

Use a Ruler

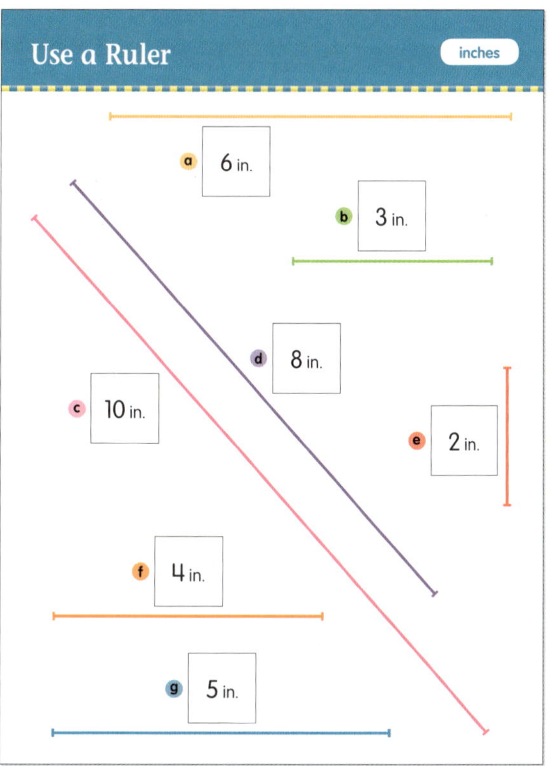

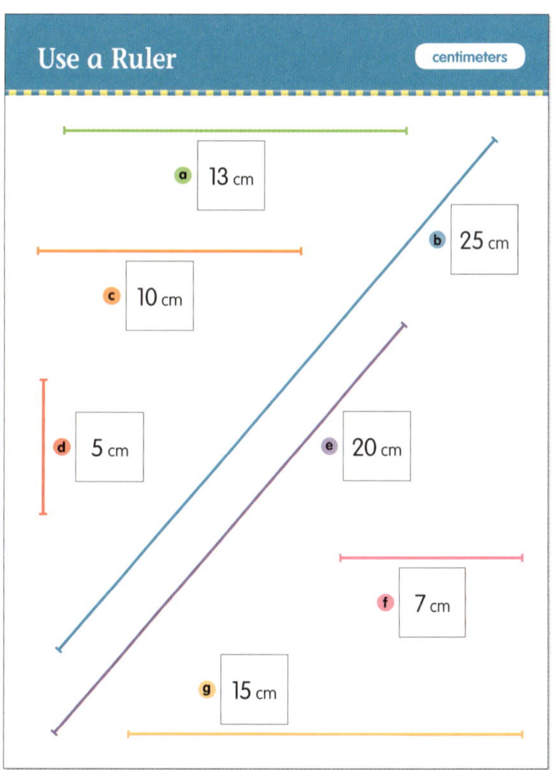

Use a Ruler

inches

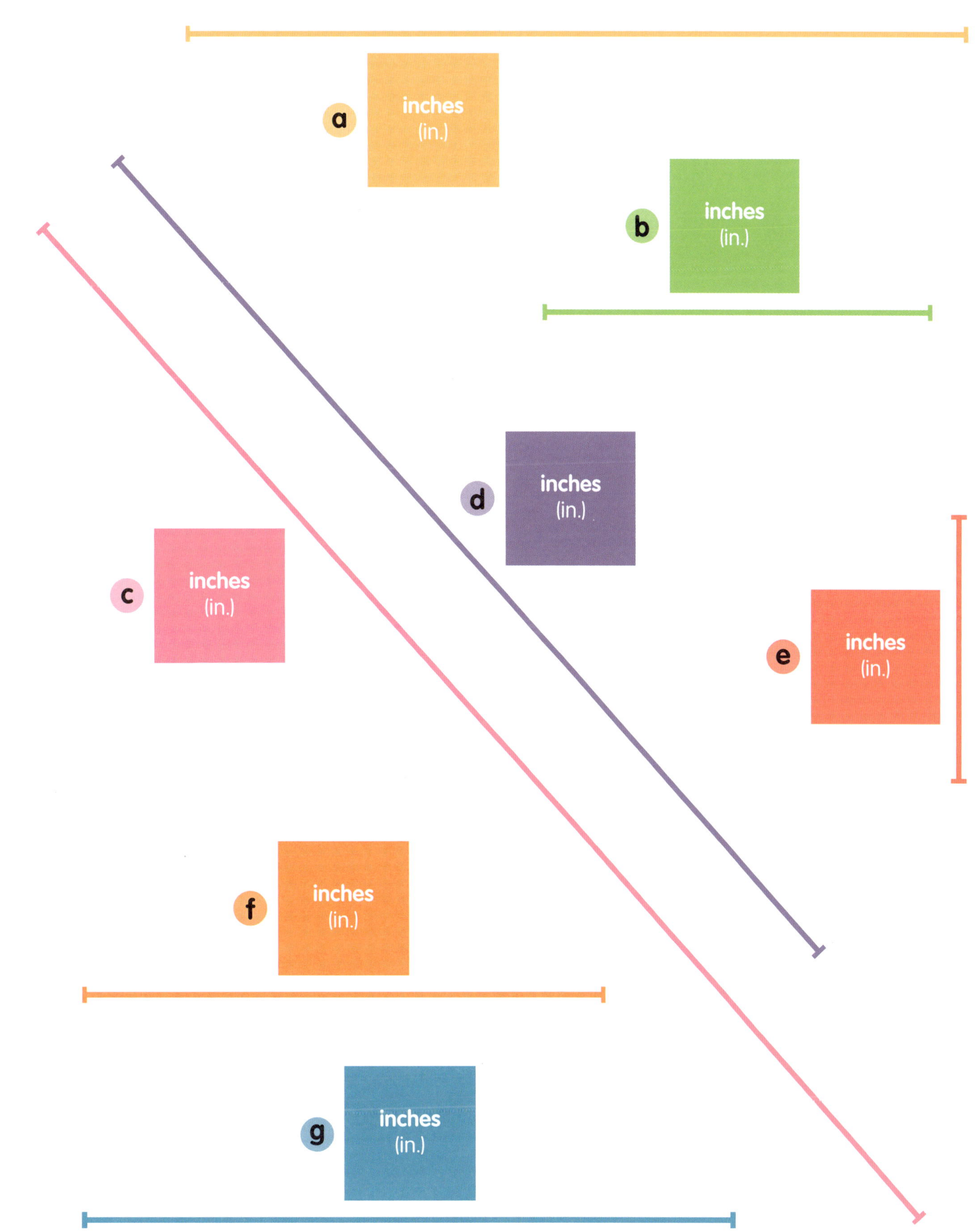

Use a Ruler

centimeters

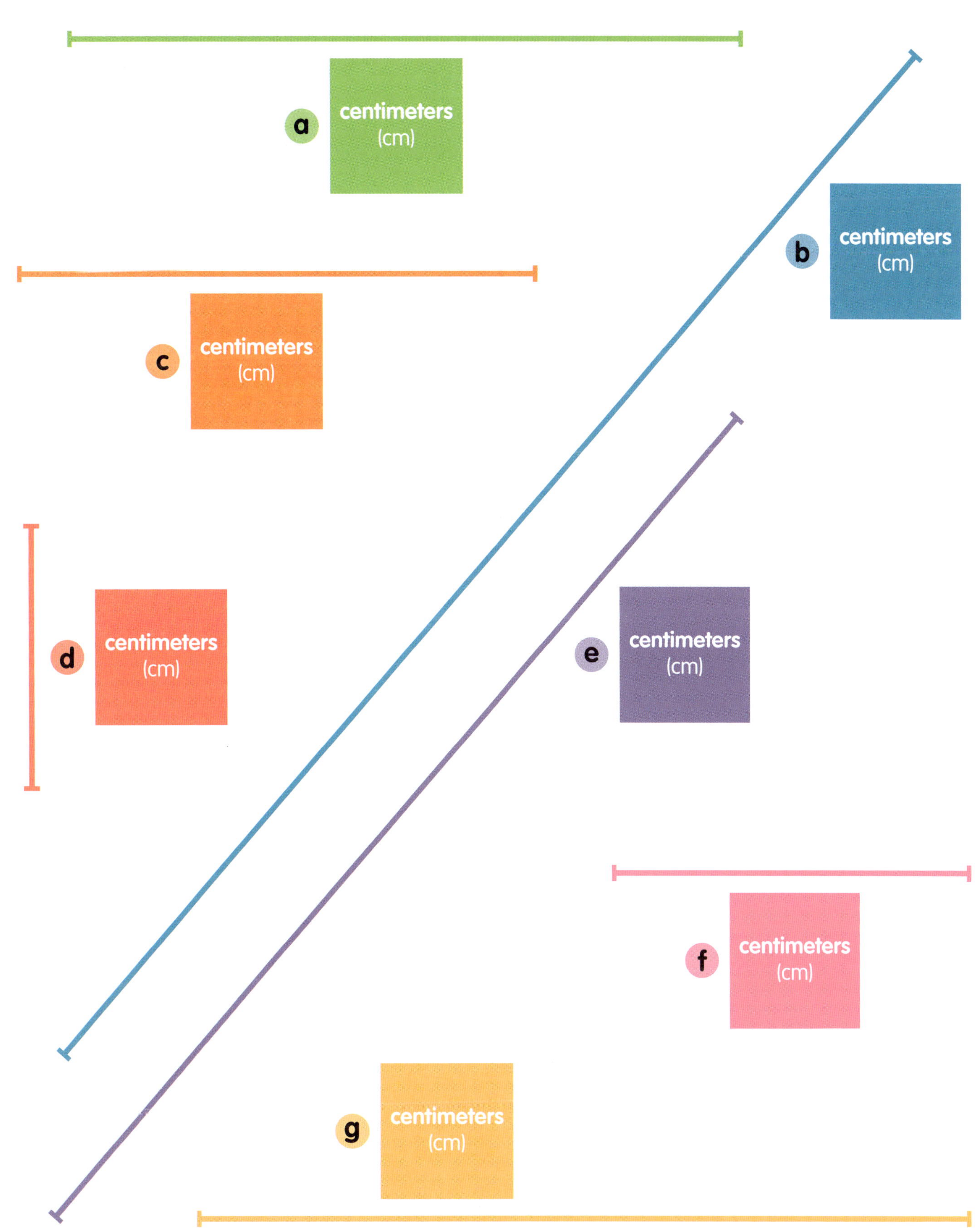

Take It to Your Seat Centers—Math • EMC 3072 • © Evan-Moor Corp.

1 in.	2 in.	3 in.	4 in.	5 in.	6 in.
7 in.	8 in.	9 in.	10 in.	11 in.	12 in.
1 cm	2 cm	3 cm	4 cm	5 cm	6 cm
7 cm	8 cm	9 cm	10 cm	11 cm	12 cm
13 cm	14 cm	15 cm	16 cm	17 cm	18 cm
19 cm	20 cm	21 cm	22 cm	23 cm	24 cm
25 cm	26 cm	27 cm	28 cm	29 cm	30 cm

Use a Ruler	Use a Ruler	Use a Ruler	Use a Ruler	Use a Ruler	Use a Ruler
EMC 3072 © Evan-Moor Corp.	EMC 3072 © Evan-Moor Corp.	EMC 3072 © Evan-Moor Corp.	EMC 3072 © Evan-Moor Corp.	EMC 3072 © Evan-Moor Corp.	EMC 3072 © Evan-Moor Corp.
Use a Ruler	Use a Ruler	Use a Ruler	Use a Ruler	Use a Ruler	Use a Ruler
EMC 3072 © Evan-Moor Corp.	EMC 3072 © Evan-Moor Corp.	EMC 3072 © Evan-Moor Corp.	EMC 3072 © Evan-Moor Corp.	EMC 3072 © Evan-Moor Corp.	EMC 3072 © Evan-Moor Corp.
Use a Ruler	Use a Ruler	Use a Ruler	Use a Ruler	Use a Ruler	Use a Ruler
EMC 3072 © Evan-Moor Corp.	EMC 3072 © Evan-Moor Corp.	EMC 3072 © Evan-Moor Corp.	EMC 3072 © Evan-Moor Corp.	EMC 3072 © Evan-Moor Corp.	EMC 3072 © Evan-Moor Corp.
Use a Ruler	Use a Ruler	Use a Ruler	Use a Ruler	Use a Ruler	Use a Ruler
EMC 3072 © Evan-Moor Corp.	EMC 3072 © Evan-Moor Corp.	EMC 3072 © Evan-Moor Corp.	EMC 3072 © Evan-Moor Corp.	EMC 3072 © Evan-Moor Corp.	EMC 3072 © Evan-Moor Corp.
Use a Ruler	Use a Ruler	Use a Ruler	Use a Ruler	Use a Ruler	Use a Ruler
EMC 3072 © Evan-Moor Corp.	EMC 3072 © Evan-Moor Corp.	EMC 3072 © Evan-Moor Corp.	EMC 3072 © Evan-Moor Corp.	EMC 3072 © Evan-Moor Corp.	EMC 3072 © Evan-Moor Corp.
Use a Ruler	Use a Ruler	Use a Ruler	Use a Ruler	Use a Ruler	Use a Ruler
EMC 3072 © Evan-Moor Corp.	EMC 3072 © Evan-Moor Corp.	EMC 3072 © Evan-Moor Corp.	EMC 3072 © Evan-Moor Corp.	EMC 3072 © Evan-Moor Corp.	EMC 3072 © Evan-Moor Corp.
Use a Ruler	Use a Ruler	Use a Ruler	Use a Ruler	Use a Ruler	Use a Ruler
EMC 3072 © Evan-Moor Corp.	EMC 3072 © Evan-Moor Corp.	EMC 3072 © Evan-Moor Corp.	EMC 3072 © Evan-Moor Corp.	EMC 3072 © Evan-Moor Corp.	EMC 3072 © Evan-Moor Corp.

Take It to Your Seat Centers

How Much Money?

Skill: Solve word problems, using coins and bills from one cent to one dollar

Steps to Follow

1. **Prepare the center.** (See page 3.)
2. **Introduce the center.** State the goal. Say: *You will find the answers to word problems by adding or subtracting dollars and cents.*
3. **Teach the skill.** Demonstrate how to use the center with individual students or small groups.
4. **Practice the skill.** Have students use the center independently or with a partner.

Contents

Written Practice........... 114

Center Cover................ 115

Answer Key.................. 117

Center Mat.................. 119

Cards.......................... 121

Name ______________________

Skill: Solve word problems, using coins and bills from one cent to one dollar

How Much Money?

Look at the mat. Write the answer for each word problem in the **amount** column below. Then write how many of each coin or bill you could use to make that amount.

	amount	$1.00 dollar	25¢ quarter	10¢ dime	5¢ nickel	1¢ penny
1						
2						
3						
4						
5						
6						
7						
8						

Take It to Your Seat Centers—Math • EMC 3072 • © Evan-Moor Corp.

Take It to Your Seat Centers

How Much Money?

Skill: Solve word problems, using coins and bills from one cent to one dollar

1. Lay out the mat and the cards.
2. Look at the list of school supplies on the mat and read how much each item costs.
3. Read each word problem. Then add or subtract to answer the question.
4. Put the card that shows the answer in the blue square.
5. Do the written practice activity.

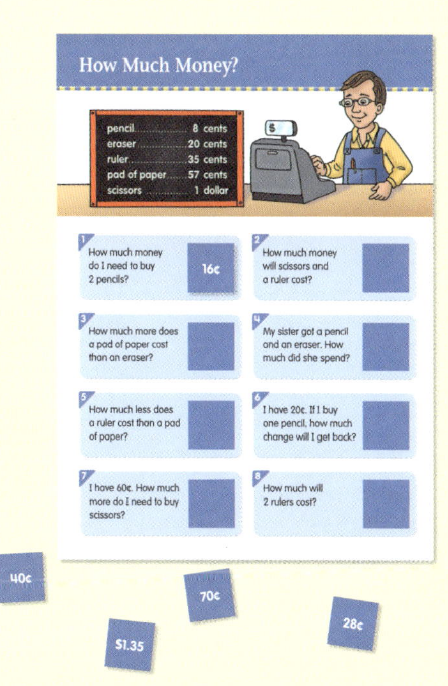

Answer Key

How Much Money?

(fold)

Written Practice

How Much Money?

Look at the mat. Write the answer for each word problem in the **amount** column below. Then write how many of each coin or bill you could use to make that amount.

	amount	$1.00 dollar	25¢ quarter	10¢ dime	5¢ nickel	1¢ penny
1	16¢					
2	$1.35					
3	37¢					
4	28¢	Answers will vary.				
5	22¢					
6	12¢					
7	40¢					
8	70¢					

Answer Key

How Much Money?

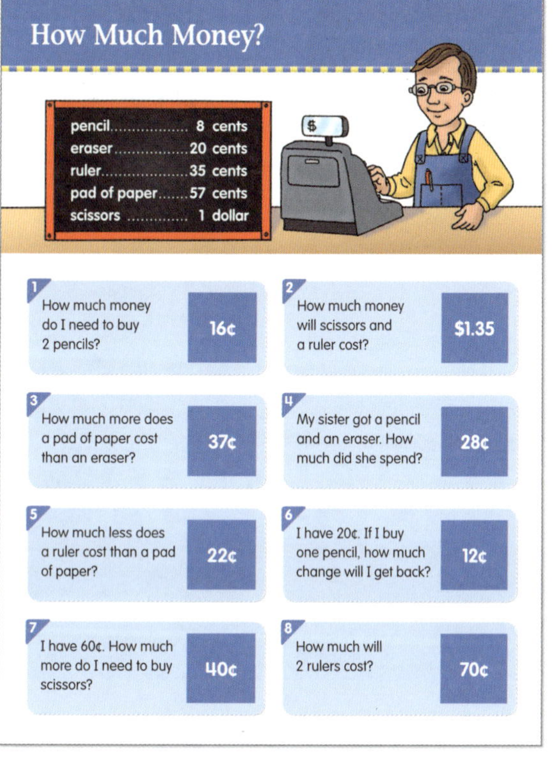

How Much Money?

pencil	8 cents
eraser	20 cents
ruler	35 cents
pad of paper	57 cents
scissors	1 dollar

1. How much money do I need to buy 2 pencils? **16¢**
2. How much money will scissors and a ruler cost? **$1.35**
3. How much more does a pad of paper cost than an eraser? **37¢**
4. My sister got a pencil and an eraser. How much did she spend? **28¢**
5. How much less does a ruler cost than a pad of paper? **22¢**
6. I have 20¢. If I buy one pencil, how much change will I get back? **12¢**
7. I have 60¢. How much more do I need to buy scissors? **40¢**
8. How much will 2 rulers cost? **70¢**

How Much Money?

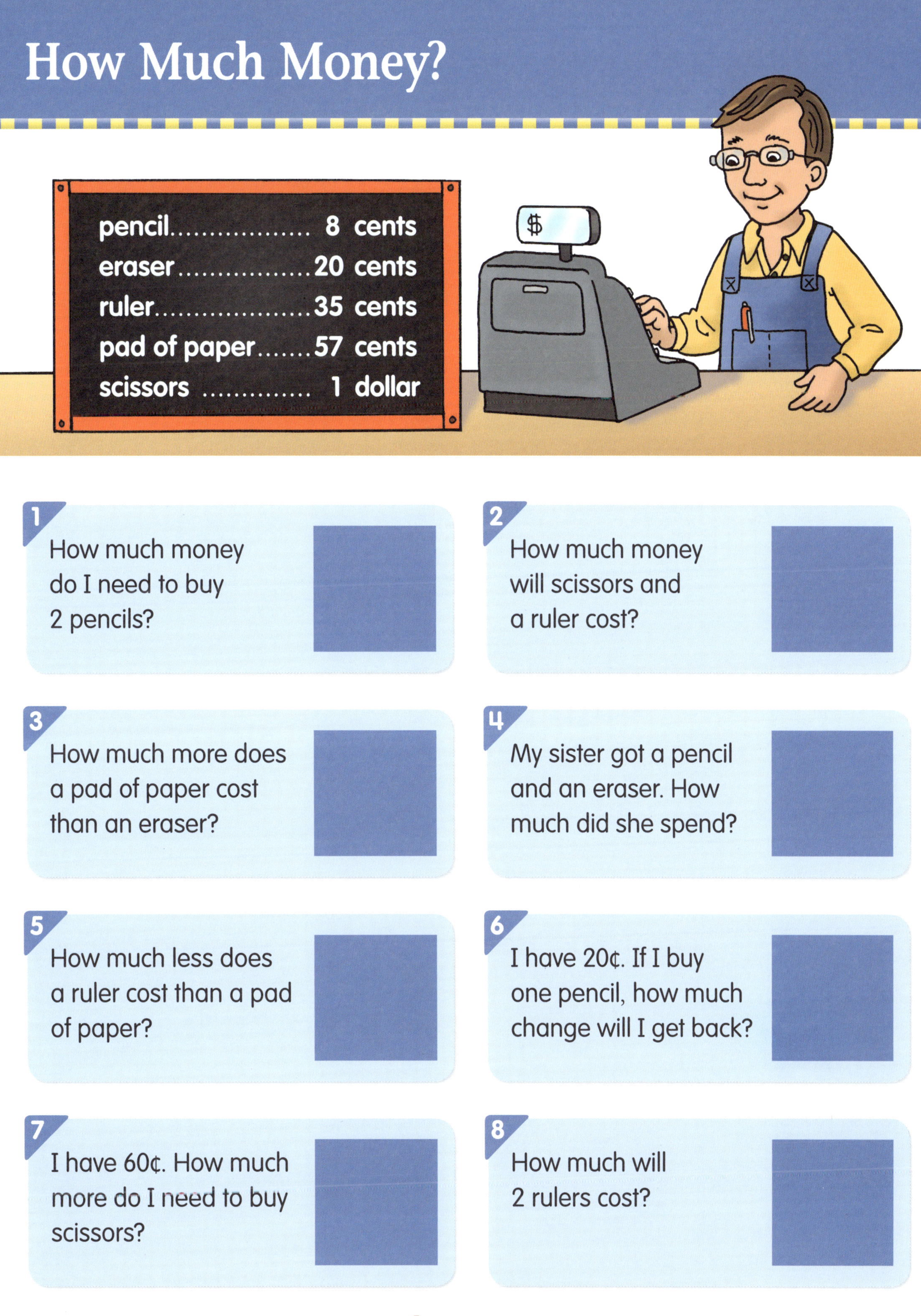

1. How much money do I need to buy 2 pencils?

2. How much money will scissors and a ruler cost?

3. How much more does a pad of paper cost than an eraser?

4. My sister got a pencil and an eraser. How much did she spend?

5. How much less does a ruler cost than a pad of paper?

6. I have 20¢. If I buy one pencil, how much change will I get back?

7. I have 60¢. How much more do I need to buy scissors?

8. How much will 2 rulers cost?

16¢	$1.35	37¢
28¢	22¢	12¢
40¢	70¢	$1.00
78¢	60¢	23¢

How Much Money?

EMC 3072

© Evan-Moor Corp.

How Much Money?

EMC 3072

© Evan-Moor Corp.

How Much Money?

EMC 3072

© Evan-Moor Corp.

How Much Money?

EMC 3072

© Evan-Moor Corp.

How Much Money?

EMC 3072

© Evan-Moor Corp.

How Much Money?

EMC 3072

© Evan-Moor Corp.

How Much Money?

EMC 3072

© Evan-Moor Corp.

How Much Money?

EMC 3072

© Evan-Moor Corp.

How Much Money?

EMC 3072

© Evan-Moor Corp.

How Much Money?

EMC 3072

© Evan-Moor Corp.

How Much Money?

EMC 3072

© Evan-Moor Corp.

How Much Money?

EMC 3072

© Evan-Moor Corp.

Take It to Your Seat Centers

What Time Is It?

Skill: Tell time to the nearest 5 minutes (analog and digital)

Steps to Follow

1. **Prepare the center.** (See page 3.)
2. **Introduce the center.** State the goal. Say: *You will find the two cards that tell the time on each clock in different ways.*
3. **Teach the skill.** Demonstrate how to use the center with individual students or small groups.
4. **Practice the skill.** Have students use the center independently or with a partner.

Contents

Written Practice............. 124
Center Cover.................. 125
Answer Key.................... 127
Center Mats 129, 131, 133
Cards 135

What Time Is It?

Write the time under each clock below.

Count by 5s.

___ ___ : ___ ___

It is quarter past.

___ ___ : ___ ___

It is half past.

___ ___ : ___ ___

Fill in the circle next to the correct time.

○ 3 o'clock
○ half past 3
○ quarter past 3

5:15

○ 5 o'clock
○ quarter to 5
○ quarter past 5

10:15

○ 10 o'clock
○ half past 10
○ quarter past 10

7:45

○ quarter to 8
○ quarter past 7
○ half past 7

○ 10 minutes to 10
○ 10 minutes to 9
○ half past 9

○ 1 o'clock
○ 10 minutes past 1
○ quarter past 1

Take It to Your Seat Centers

What Time Is It?

Skill: Tell time to the nearest 5 minutes

There are five minutes between each number on a clock.

1. Lay out the mats and sort the cards by color.
2. Look at each clock on the mats. Can you tell what time it is?
3. Find the gold card and the blue card that show the time in different ways.
4. Put the cards in the boxes next to the clock.
5. Do the written practice activity.

Answer Key

What Time Is It?

(fold)

Written Practice

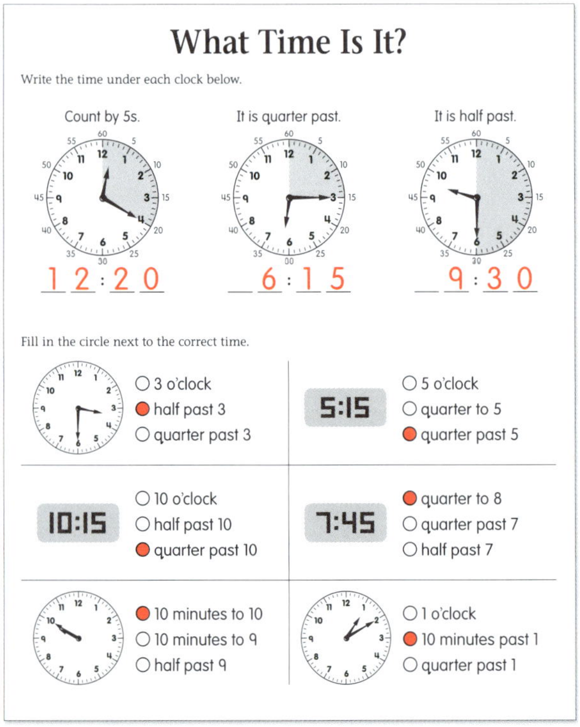

What Time Is It?

Write the time under each clock below.

Count by 5s.	It is quarter past.	It is half past.
12:20	6:15	9:30

Fill in the circle next to the correct time.

Clock	Choices
(clock)	○ 3 o'clock ● half past 3 ○ quarter past 3
5:15	○ 5 o'clock ○ quarter to 5 ● quarter past 5
10:15	○ 10 o'clock ○ half past 10 ● quarter past 10
7:45	● quarter to 8 ○ quarter past 7 ○ half past 7
(clock)	● 10 minutes to 10 ○ 10 minutes to 9 ○ half past 9
(clock)	○ 1 o'clock ● 10 minutes past 1 ○ quarter past 1

Answer Key

What Time Is It?

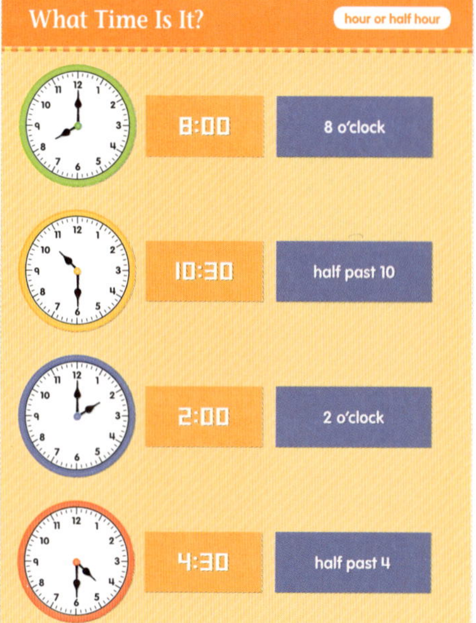

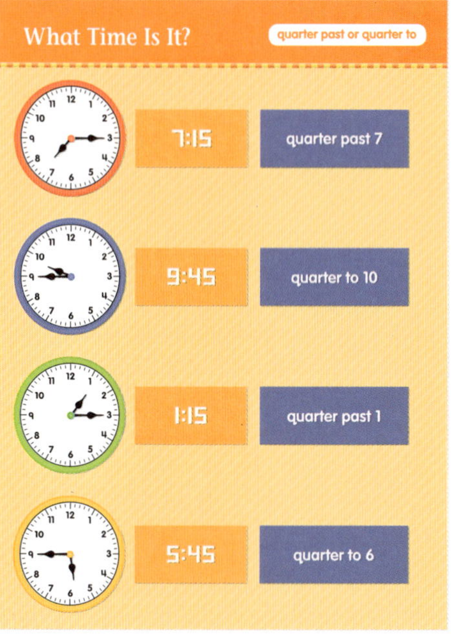

What Time Is It?

nearest 5 minutes

12:10 — 10 minutes past 12

3:05 — 5 minutes past 3

6:40 — 20 minutes to 7

11:55 — 5 minutes to 12

What Time Is It?

hour or half hour

What Time Is It?

quarter past or quarter to

 © Evan-Moor Corp.

What Time Is It?

nearest 5 minutes

8:00	10:30	2:00	4:30
7:15	9:45	1:15	5:45
12:10	3:05	6:40	11:55

8 o'clock	quarter past 7	10 minutes past 12
half past 10	quarter to 10	20 minutes to 7
2 o'clock	quarter past 1	5 minutes past 3
half past 4	quarter to 6	5 minutes to 12

What Time Is It?
EMC 3072
© Evan-Moor Corp.

What Time Is It?
EMC 3072
© Evan-Moor Corp.

What Time Is It?
EMC 3072
© Evan-Moor Corp.

What Time Is It?
EMC 3072
© Evan-Moor Corp.

What Time Is It?
EMC 3072
© Evan-Moor Corp.

What Time Is It?
EMC 3072
© Evan-Moor Corp.

What Time Is It?
EMC 3072
© Evan-Moor Corp.

What Time Is It?
EMC 3072
© Evan-Moor Corp.

What Time Is It?
EMC 3072
© Evan-Moor Corp.

What Time Is It?
EMC 3072
© Evan-Moor Corp.

What Time Is It?
EMC 3072
© Evan-Moor Corp.

What Time Is It?
EMC 3072
© Evan-Moor Corp.

What Time Is It?
EMC 3072
© Evan-Moor Corp.

What Time Is It?
EMC 3072
© Evan-Moor Corp.

What Time Is It?
EMC 3072
© Evan-Moor Corp.

What Time Is It?
EMC 3072
© Evan-Moor Corp.

What Time Is It?
EMC 3072
© Evan-Moor Corp.

What Time Is It?
EMC 3072
© Evan-Moor Corp.

What Time Is It?
EMC 3072
© Evan-Moor Corp.

What Time Is It?
EMC 3072
© Evan-Moor Corp.

What Time Is It?
EMC 3072
© Evan-Moor Corp.

What Time Is It?
EMC 3072
© Evan-Moor Corp.

What Time Is It?
EMC 3072
© Evan-Moor Corp.

What Time Is It?
EMC 3072
© Evan-Moor Corp.

Take It to Your Seat Centers

Graphs and Data

Skill: Read and use data on bar graphs and picture graphs

Steps to Follow

1. **Prepare the center.** (See page 3.)
2. **Introduce the center.** State the goal. Say: *You will read the graph on each mat and answer questions about the data it shows.*
3. **Teach the skill.** Demonstrate how to use the center with individual students or small groups.
4. **Practice the skill.** Have students use the center independently or with a partner.

Contents

Written Practice........... 138
Center Cover................ 139
Answer Key.................. 141
Center Mats 143, 145
Cards 147

Name ______________________

Skill: Read and use data on bar graphs and picture graphs

Graphs and Data

Use the graph below to answer the questions.

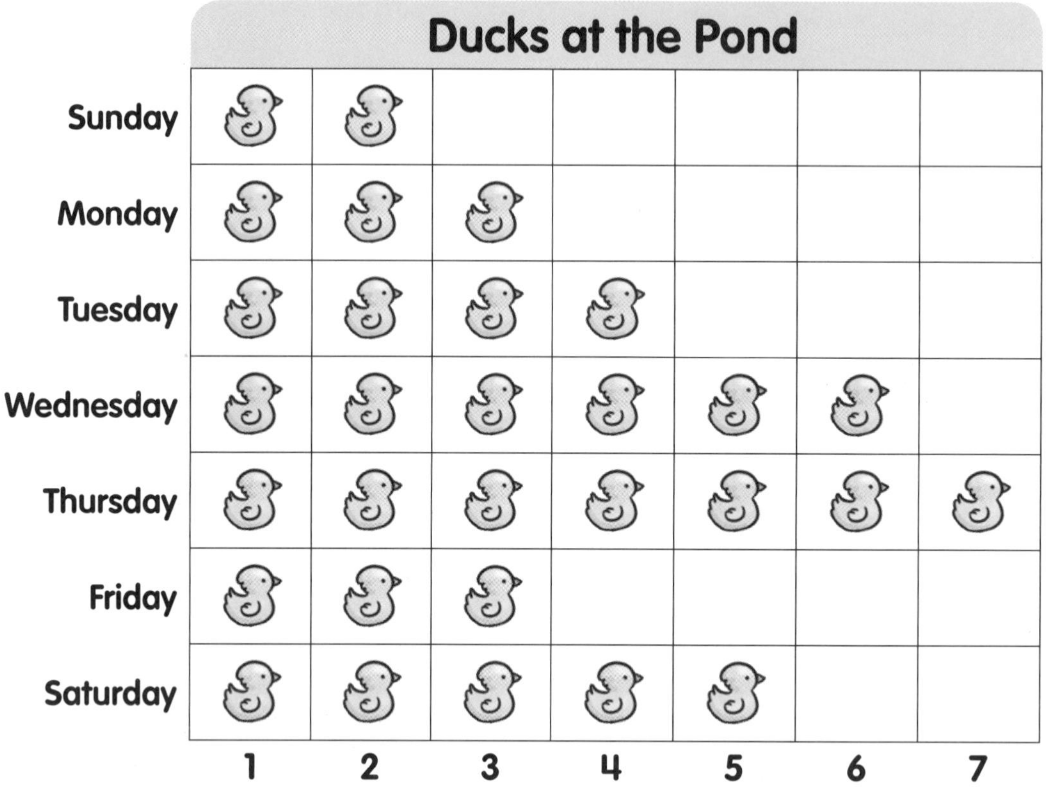

Ducks at the Pond

Sunday
Monday
Tuesday
Wednesday
Thursday
Friday
Saturday

1 2 3 4 5 6 7

1. How many ducks were at the pond on Sunday? ________

2. How many more ducks were at the pond on Saturday than on Friday? ________

3. How many ducks altogether were at the pond from Monday through Wednesday? ________

4. Which day had the most ducks? ______________________

5. Which days had the same number of ducks?

____

 © Evan-Moor Corp.

Take It to Your Seat Centers

Graphs and Data

Skill: Read and use data on bar graphs and picture graphs

Graphs can use bars or pictures to show how many.

1. Lay out the mats and the cards.
2. Look at the graph on each mat and read the questions.
3. Use the graph to find the answer to each question.
4. Find the card that shows the answer and put it in the colored square.
5. Do the written practice activity.

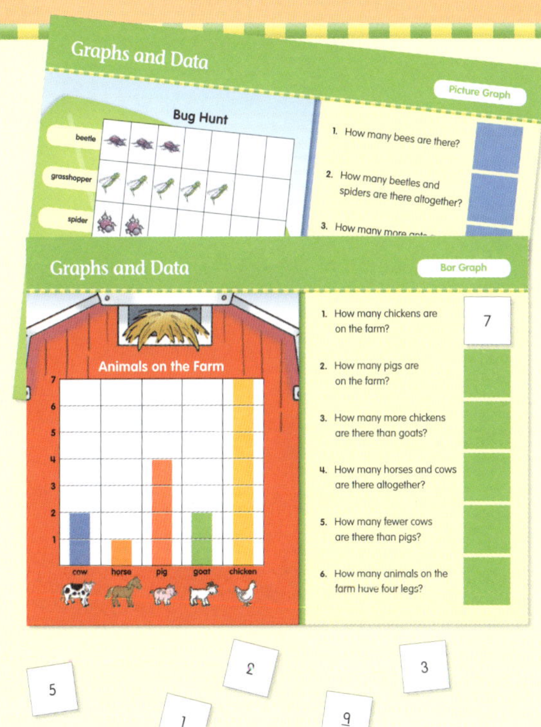

Answer Key

Graphs and Data

(fold)

Written Practice

Graphs and Data

Use the graph below to answer the questions.

Ducks at the Pond

Sunday							
Monday							
Tuesday							
Wednesday							
Thursday							
Friday							
Saturday							
	1	2	3	4	5	6	7

1. How many ducks were at the pond on Sunday? 2
2. How many more ducks were at the pond on Saturday than on Friday? 2
3. How many ducks altogether were at the pond from Monday through Wednesday? 13
4. Which day had the most ducks? Thursday
5. Which days had the same number of ducks? Monday and Friday

Answer Key

Graphs and Data

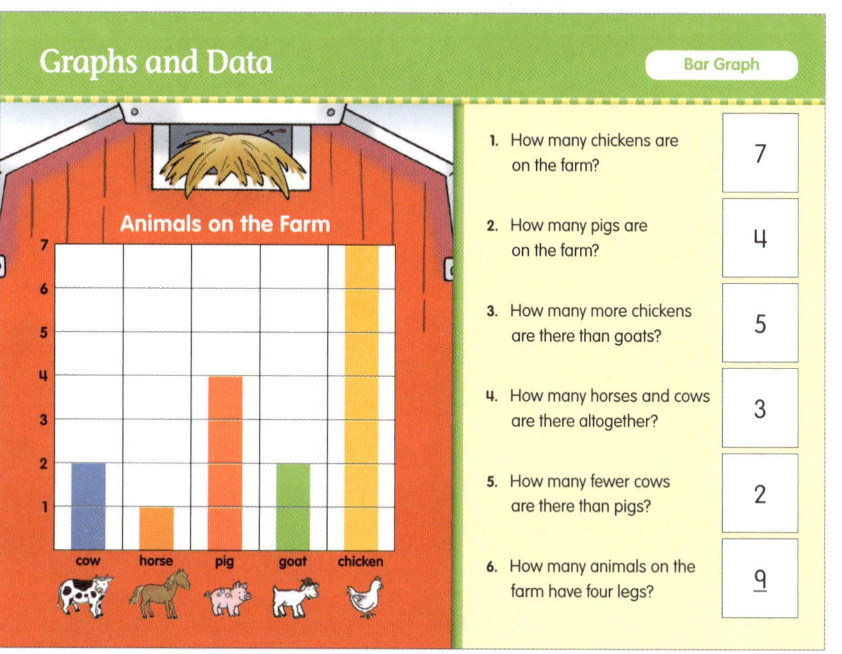

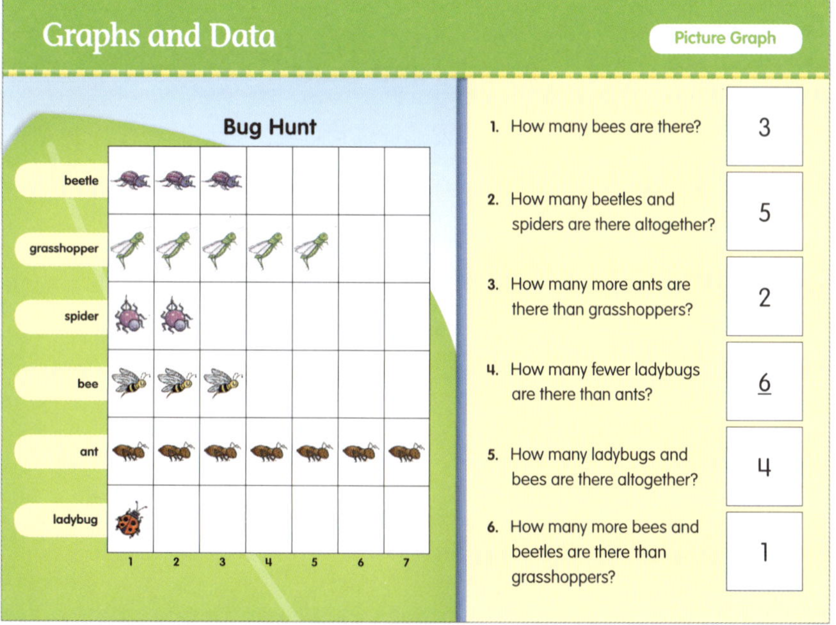

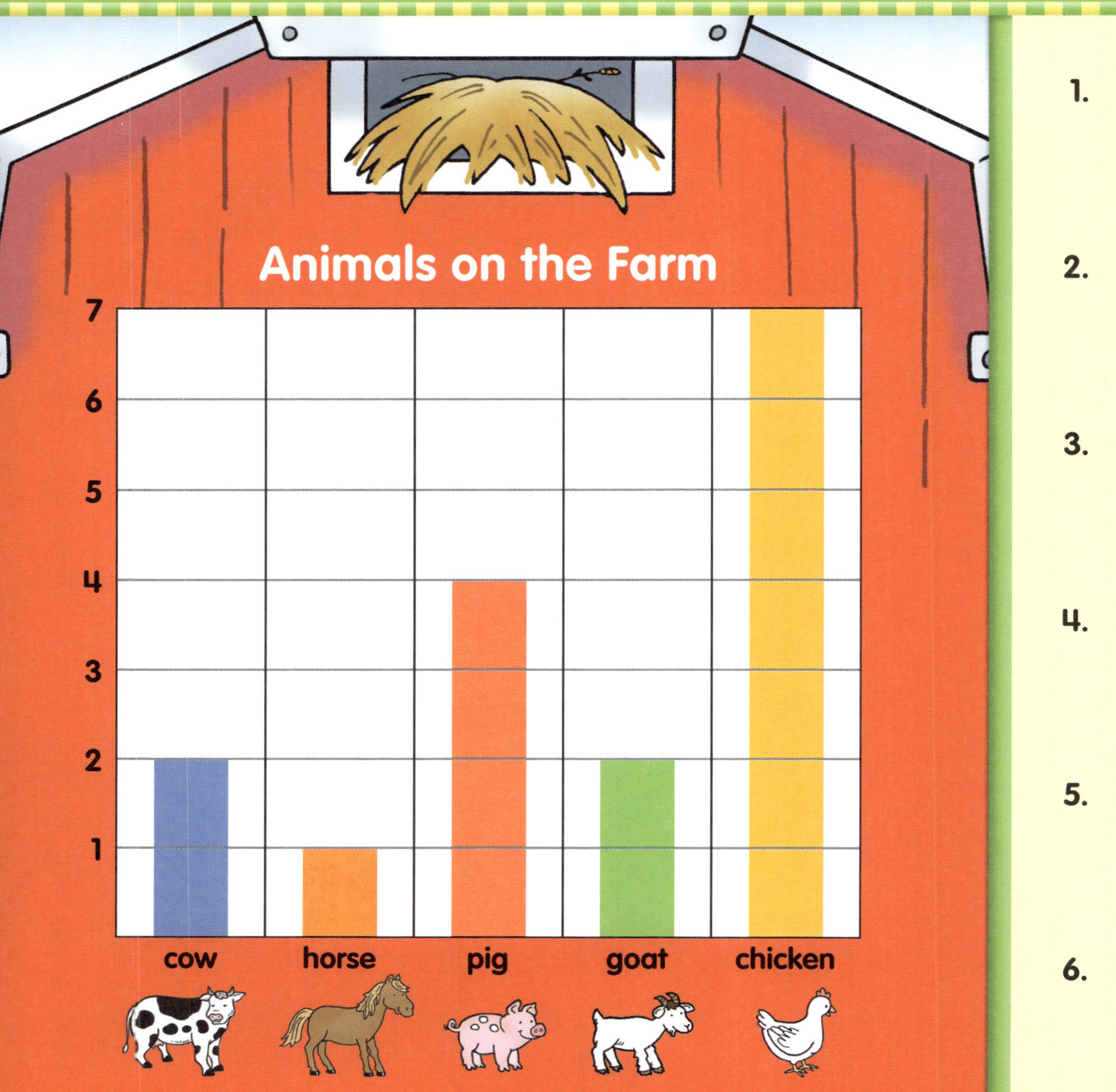

1. How many chickens are on the farm?
2. How many pigs are on the farm?
3. How many more chickens are there than goats?
4. How many horses and cows are there altogether?
5. How many fewer cows are there than pigs?
6. How many animals on the farm have four legs?

1. How many bees are there?

2. How many beetles and spiders are there altogether?

3. How many more ants are there than grasshoppers?

4. How many fewer ladybugs are there than ants?

5. How many ladybugs and bees are there altogether?

6. How many more bees and beetles are there than grasshoppers?

1	1	2	2	3
3	4	4	5	5
6	6	7	7	8
8	9	9	10	10

Graphs and Data EMC 3072 © Evan-Moor Corp.	Graphs and Data EMC 3072 © Evan-Moor Corp.	Graphs and Data EMC 3072 © Evan-Moor Corp.	Graphs and Data EMC 3072 © Evan-Moor Corp.	Graphs and Data EMC 3072 © Evan-Moor Corp.
Graphs and Data EMC 3072 © Evan-Moor Corp.	Graphs and Data EMC 3072 © Evan-Moor Corp.	Graphs and Data EMC 3072 © Evan-Moor Corp.	Graphs and Data EMC 3072 © Evan-Moor Corp.	Graphs and Data EMC 3072 © Evan-Moor Corp.
Graphs and Data EMC 3072 © Evan-Moor Corp.	Graphs and Data EMC 3072 © Evan-Moor Corp.	Graphs and Data EMC 3072 © Evan-Moor Corp.	Graphs and Data EMC 3072 © Evan-Moor Corp.	Graphs and Data EMC 3072 © Evan-Moor Corp.
Graphs and Data EMC 3072 © Evan-Moor Corp.	Graphs and Data EMC 3072 © Evan-Moor Corp.	Graphs and Data EMC 3072 © Evan-Moor Corp.	Graphs and Data EMC 3072 © Evan-Moor Corp.	Graphs and Data EMC 3072 © Evan-Moor Corp.

Take It to Your Seat Centers

Shapes and Fractions

Skill: Identify fractional parts ($\frac{1}{2}$, $\frac{1}{3}$, $\frac{1}{4}$) of shapes with 3 to 6 sides

Steps to Follow

1. **Prepare the center.** (See page 3.)
2. **Introduce the center.** State the goal. Say: *You will read each description on the mat and find the card that shows that shape and fraction.*
3. **Teach the skill.** Demonstrate how to use the center with individual students or small groups.
4. **Practice the skill.** Have students use the center independently or with a partner.

Contents

Written Practice........... 150
Center Cover................ 151
Answer Key.................. 153
Center Mats......... 155, 157
Cards.......................... 159

Name ______________________

Skill: Identify fractional parts of shapes with 3 to 6 sides

Shapes and Fractions

Write the fraction that shows how much of each shape is shaded.

1. 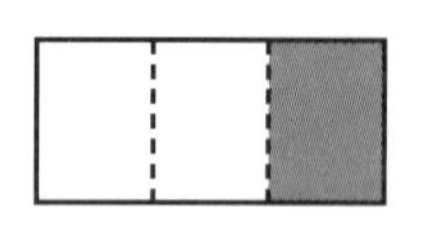 $\frac{1}{3}$

2. 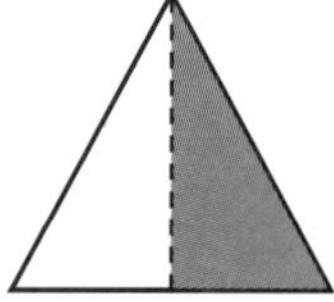 ___

3. 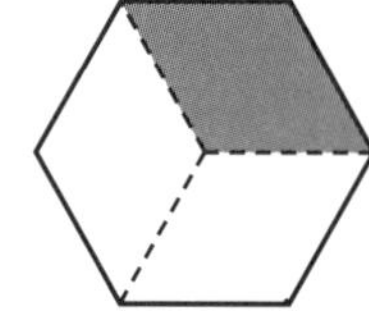 ___

4. 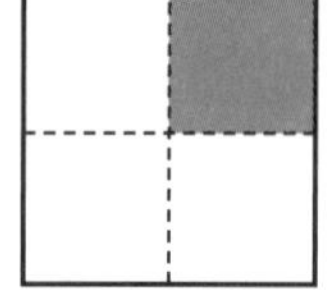 ___

5. 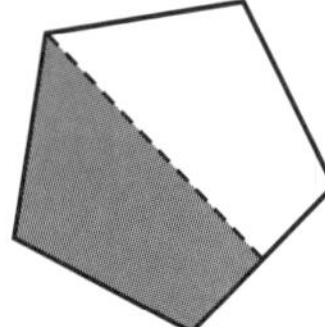 ___

6. 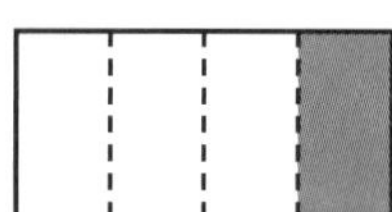 ___

7. 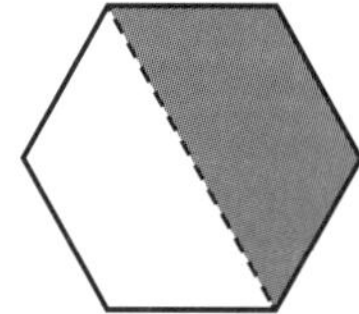 ___

8.  ___

9.  ___

10. 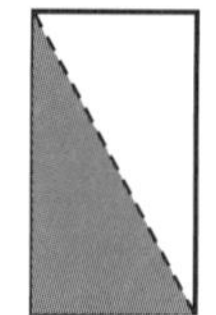 ___

11. 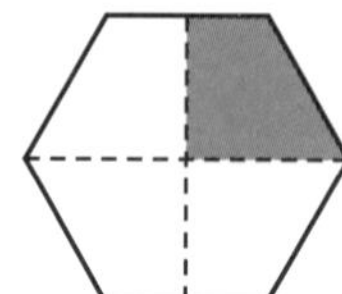 ___

12. 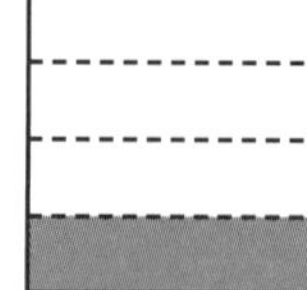 ___

13. 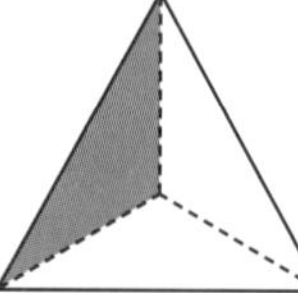 ___

14. 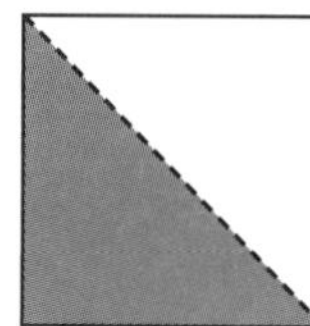 ___

Take It to Your Seat Centers—Math • EMC 3072 • © Evan-Moor Corp.

Take It to Your Seat Centers

Shapes and Fractions

Skill: Identify fractional parts of shapes with 3 to 6 sides

You can divide shapes into parts.

triangle

rectangle

square

pentagon

hexagon

1. Lay out the mats and the cards.
2. Read each description on the mats.
3. Find the card that shows the shape and the fraction described and put it in the yellow square.
4. Do the written practice activity.

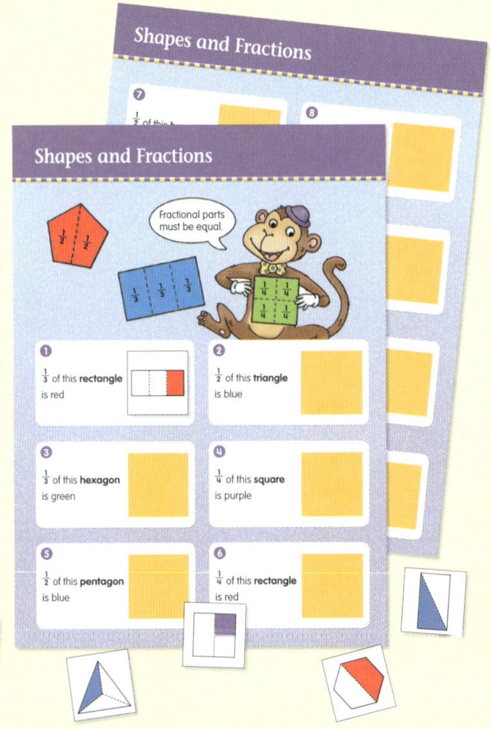

(fold)

Answer Key

Shapes and Fractions

Written Practice

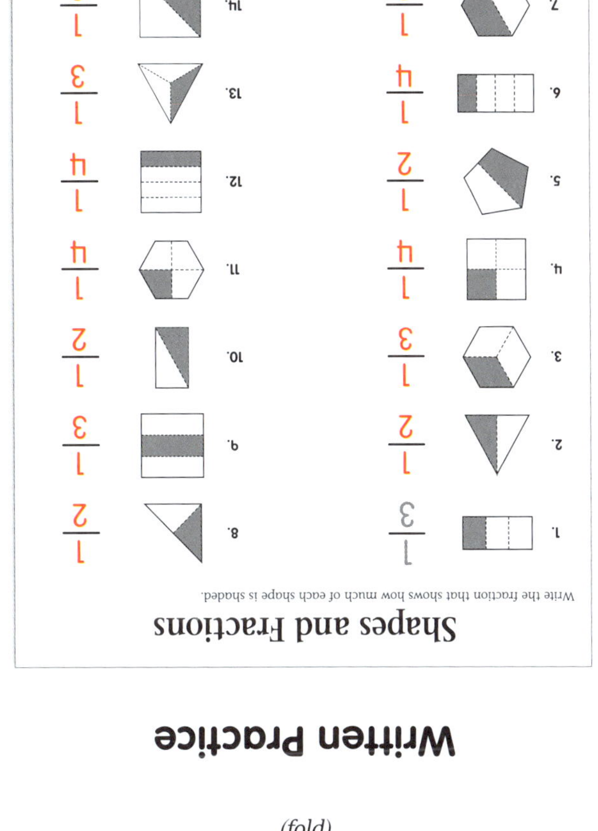
Shapes and Fractions

Write the fraction that shows how much of each shape is shaded.

1. $\frac{1}{3}$
2. $\frac{1}{2}$
3. $\frac{1}{3}$
4. $\frac{1}{4}$
5. $\frac{1}{2}$
6. $\frac{1}{4}$
7. $\frac{1}{2}$
8. $\frac{1}{2}$
9. $\frac{1}{3}$
10. $\frac{1}{2}$
11. $\frac{1}{4}$
12. $\frac{1}{4}$
13. $\frac{1}{3}$
14. $\frac{1}{2}$

Answer Key

Shapes and Fractions

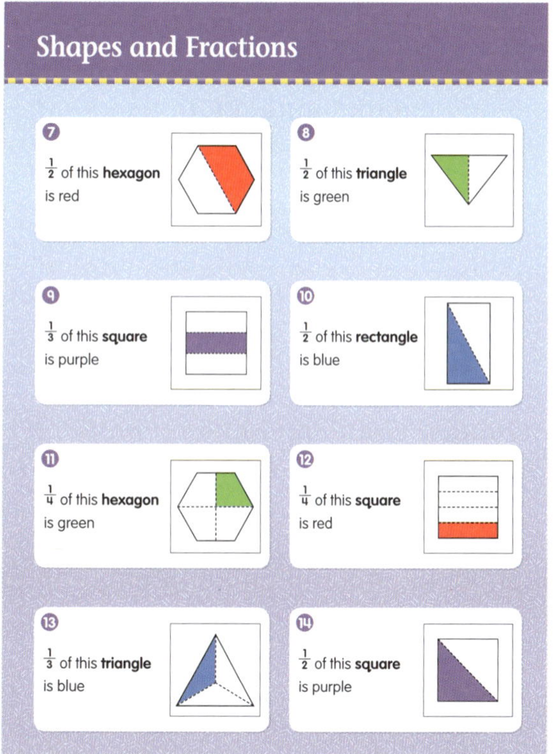

Shapes and Fractions

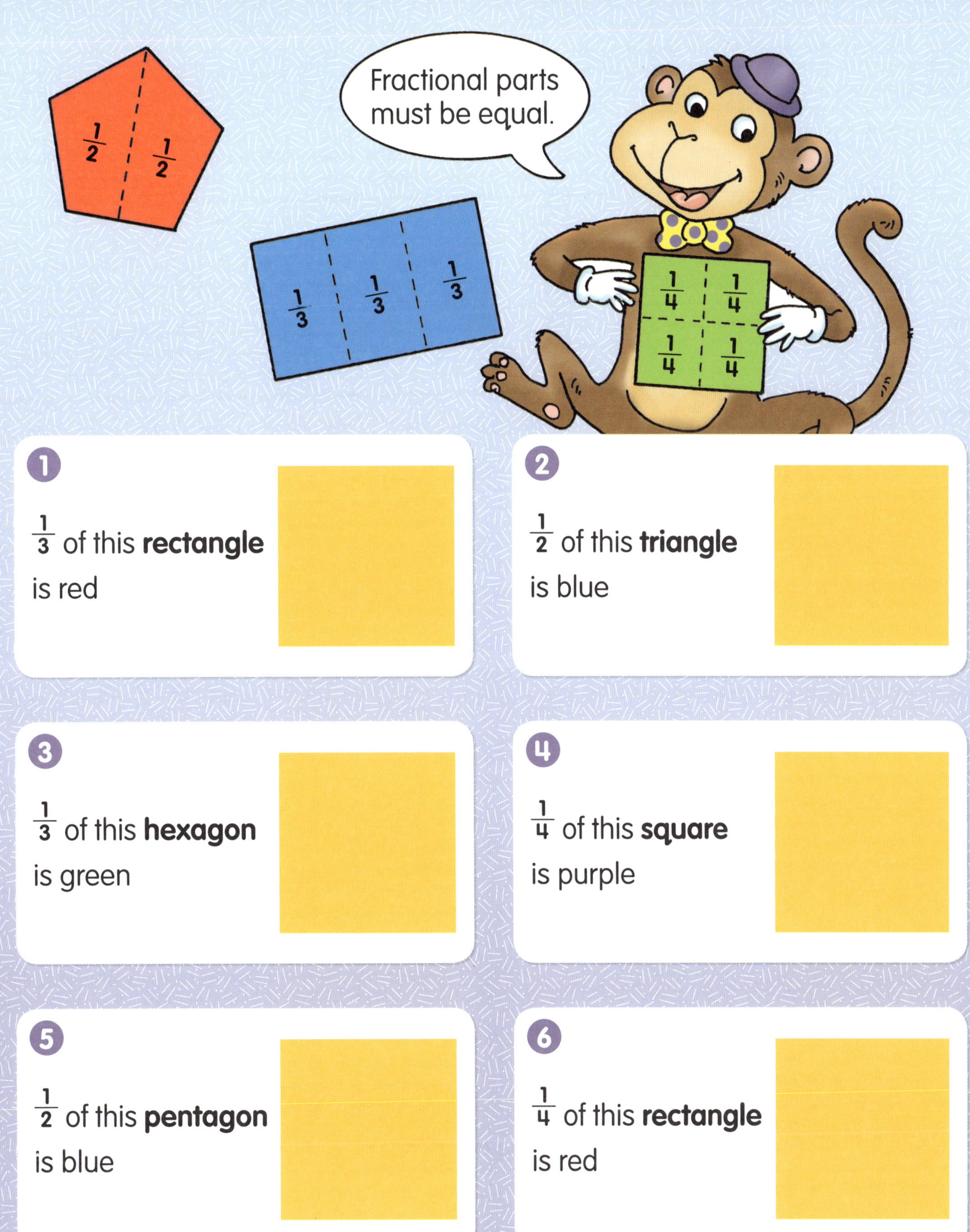

1. $\frac{1}{3}$ of this **rectangle** is red

2. $\frac{1}{2}$ of this **triangle** is blue

3. $\frac{1}{3}$ of this **hexagon** is green

4. $\frac{1}{4}$ of this **square** is purple

5. $\frac{1}{2}$ of this **pentagon** is blue

6. $\frac{1}{4}$ of this **rectangle** is red

Shapes and Fractions

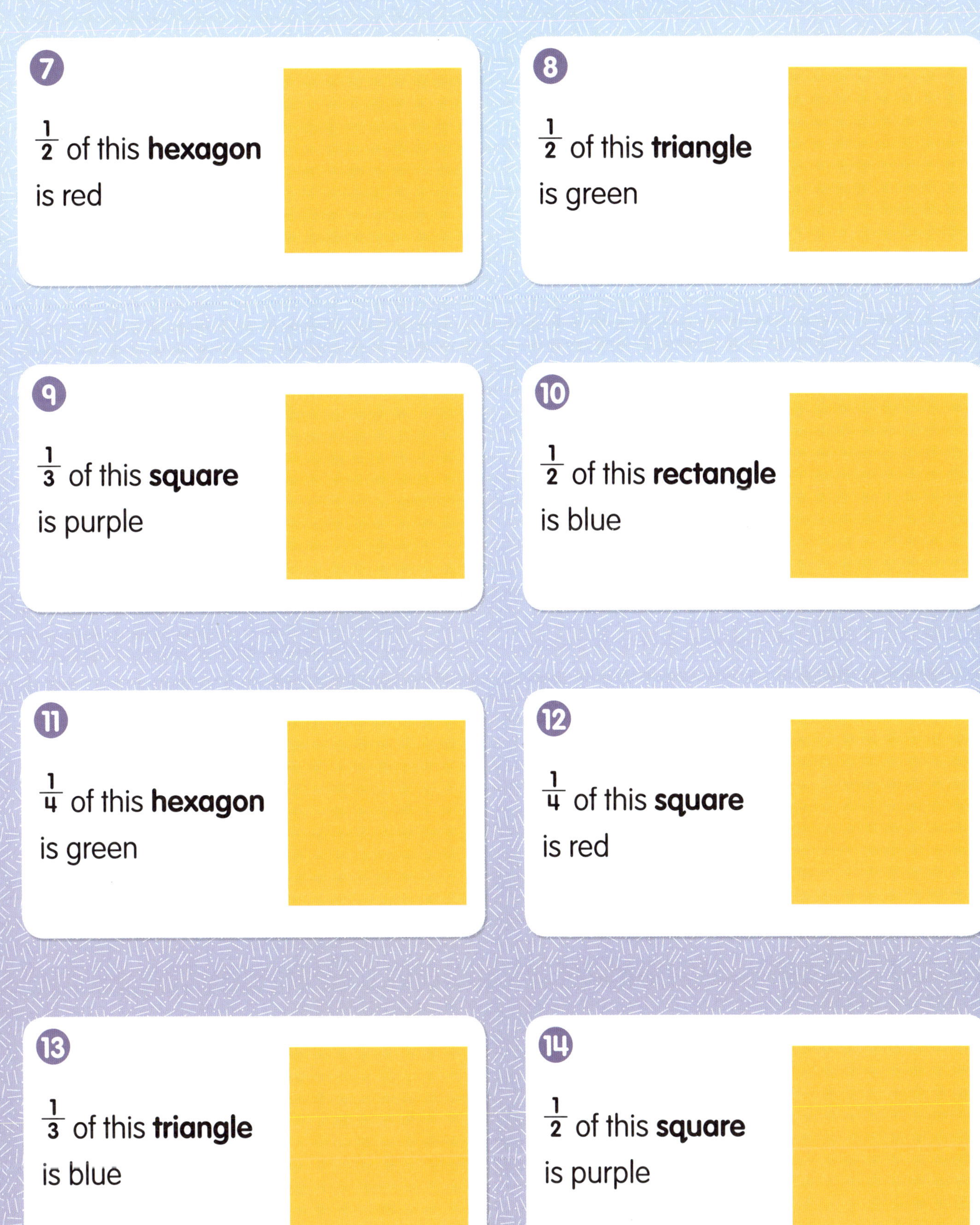

7

$\frac{1}{2}$ of this **hexagon** is red

8

$\frac{1}{2}$ of this **triangle** is green

9

$\frac{1}{3}$ of this **square** is purple

10

$\frac{1}{2}$ of this **rectangle** is blue

11

$\frac{1}{4}$ of this **hexagon** is green

12

$\frac{1}{4}$ of this **square** is red

13

$\frac{1}{3}$ of this **triangle** is blue

14

$\frac{1}{2}$ of this **square** is purple

Take It to Your Seat Centers—Math • EMC 3072 • © Evan-Moor Corp.

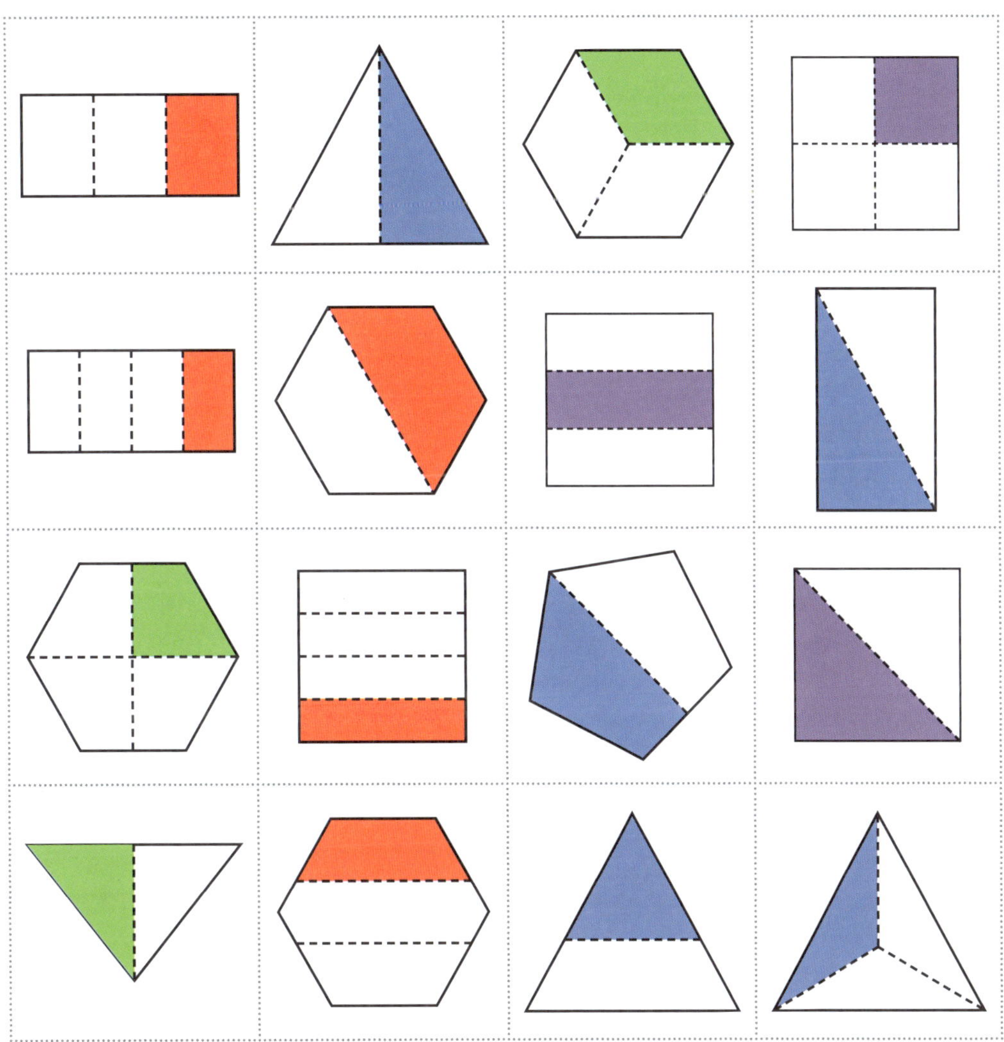

Shapes and Fractions

EMC 3072

© Evan-Moor Corp.

Shapes and Fractions

EMC 3072

© Evan-Moor Corp.

Shapes and Fractions

EMC 3072

© Evan-Moor Corp.

Shapes and Fractions

EMC 3072

© Evan-Moor Corp.

Shapes and Fractions

EMC 3072

© Evan-Moor Corp.

Shapes and Fractions

EMC 3072

© Evan-Moor Corp.

Shapes and Fractions

EMC 3072

© Evan-Moor Corp.

Shapes and Fractions

EMC 3072

© Evan-Moor Corp.

Shapes and Fractions

EMC 3072

© Evan-Moor Corp.

Shapes and Fractions

EMC 3072

© Evan-Moor Corp.

Shapes and Fractions

EMC 3072

© Evan-Moor Corp.

Shapes and Fractions

EMC 3072

© Evan-Moor Corp.

Shapes and Fractions

EMC 3072

© Evan-Moor Corp.

Shapes and Fractions

EMC 3072

© Evan-Moor Corp.

Shapes and Fractions

EMC 3072

© Evan-Moor Corp.

Shapes and Fractions

EMC 3072

© Evan-Moor Corp.